T0092677

Electricity

Teruo Matsushita

Electricity

Electromagnetism and Electric Circuits

 Springer

Teruo Matsushita
Fukuoka, Japan

ISBN 978-3-031-44001-4   ISBN 978-3-031-44002-1   (eBook)
https://doi.org/10.1007/978-3-031-44002-1

This Springer imprint is published by the registered company Springer Nature Switzerland AG
The registered company address is: Gewerbestrasse 11, 6330 Cham, Switzerland

Paper in this product is recyclable.

Preface

The electric circuit is an essential topic in electric engineering and electronics. Its behavior is based on electromagnetism and gives the basis for electronic circuit. So, it is especially important for students who learn hardware systems such as large-scale integration (LSI) that are composed of electronic circuits. It is also useful for students in other fields who learn electric control systems, mechanical systems, artificial intelligence, or even bioinformatics. This book is written for such students in other fields who need some knowledge of electric circuits.

In this book students will learn first electromagnetism and then move to electric circuits. Electromagnetism is the basis for exactly understanding the features of circuit elements such as the resistor, capacitor, and coil that compose electric circuits. In particular, an understanding of electromagnetism is essential to quantitatively determine the resistance, capacitance, and inductance of circuit elements. The relationship between the current and voltage in each element is also obtained by this study. Electromagnetic induction, which is a basic principle of the electric generator and the transformer, is also a key phenomenon of electromagnetism. In addition, Kirchhoff's laws, which are extremely important laws in electric circuits, are directly derived from electromagnetism. A careful argument is needed to apply Kirchhoff's laws to AC circuits from viewpoint of electromagnetism. In this book, the treatments of electromagnetism and electric circuits are compactly packaged, and the correlation between them is described in detail. This is the largest advantage of this book. Each volume of electromagnetism and electric circuits is suitable for lectures over a semester. Thus, it is effective to give lectures on both over a year. It is also possible to give lectures on each part independently.

Characteristic points on electromagnetism and electric circuits in this book are introduced. The E-B analogy is now commonly used in electromagnetism. In particular, the superconductor is introduced as a kind of magnetic material to strengthen the E-B analogy: Magnetic phenomena around superconductors, which show that $B = 0$, are compared with electric phenomena around conductors, which show that $E = 0$. This contributes to strengthening the analogy between electricity and magnetism with the correspondence between electric phenomena in dielectric materials and magnetic phenomena in magnetic materials. In addition, the introduction of superconductors is expected to be useful to study superconducting circuits that will be realized in the future. From a technical viewpoint, integral formulae, which are mostly used to solve problems, are adopted to assist the readers in easily gaining understanding. The deferential formulae that are useful to characterize the field and potential are described in Appendix.

In the treatment of electric circuits, we first learn about DC electric circuits composed of resistors and DC voltage sources. Here, the intuitive and analytic solution methods are compared to strengthen the importance of finding the essential point. Then, the transient and steady-state responses of AC electric circuits are treated. In the transient response we learn about the energy stored or dissipated in each circuit element. After that, AC circuit theory is introduced for describing the AC steady-state phenomena in circuits. In particular, we learn the characteristics of various types of filter from the frequency response. In addition, the advantages of using complex electric power are also introduced. In electric circuits, the electromotive force generated by electromagnetic induction is not treated as an electromotive force, but as a voltage drop. This treatment, which is an extension of Kirchhoff's second law to AC circuits, is explained in detail from the viewpoint of electromagnetism. Finally, theoretical treatments such as the branch current method, closed current method, and node potential method are introduced to show the validity of the solution methods that have been used in the book. Not only analytic methods, but intuitive methods are also used to solve some problems, since it is a good exercise to find the essential points in problems.

Many examples and exercises are arranged in each chapter to yield a deeper understanding. The answers to exercises are explained in detail at the end of the book. It is highly desirable to use them effectively. Finally, the author would like to acknowledge Dr. Tania M. Silver at Wollongong University for editing the English in the book.

Fukuoka, Japan   Teruo Matsushita

Contents

Chapter 1
Electric Phenomena in Vacuum

Abstract This chapter covers electric phenomena in vacuum due to electric charges. The Coulomb force works between electric charges. This electric reaction is considered to be caused by an electrical distortion in space that is produced by other electric charges, and this distortion is called the electric field. The local electric field is described by Coulomb's law. On the other hand, the global relationship between the electric charge and the electric field is described by Guass's law. This law is sometimes useful to determine the electric field strength. The electric field can be regarded as caused by an electric potential, which is equal to the mechanical work to carry a unit electric charge under the reaction by the electric field. The electric potential is derived by a curvilinear integral of the electric field. The electrostatic energy is equal to the product of the electric charge and the electric potential.

1.1 Electric Charge in Vacuum

The substance that causes various electric phenomena is **electric charge**, and innumerable protons with positive electric charge and electrons with negative electric charge exist in materials. The effects of these electric charges are cancelled and electrically neutral states are the result in most cases. When this balance is broken, electric phenomena appear. The amount of electric charge that one proton has is the **elementary electric charge**:

$$e = 1.602176634 \times 10^{-19} \text{C}, \tag{1.1}$$

where the unit [C] is coulomb. The amount of electric charge of electron is $-e$.

There are two kinds of electric charge that cause electric phenomena: One is **true electric charge**, which can be transferred outside a substance, and the other is **polarization charge**, which cannot be transferred outside due to locally binding around nuclei. The former appears on the surface of a conductor, and its flow gives a current, while the latter appears on the surface of a dielectric material.

The **principle of conservation of charge** holds, i.e., the total amount of electric charge is constant in a closed system. Even when positive and negative charges cancel

T. Matsushita, *Electricity*,
https://doi.org/10.1007/978-3-031-44002-1_1

each other, the algebraic sum of electric charge is unchanged. For example, when a piece of amber is scrubbed with a silk cloth, positive charge appears on the amber, and the corresponding negative charge appears on the silk cloth. The total sum of the electric charge is zero and is the same as before scrubbing. When the system is not closed and the total sum of electric charge changes, movement of electric charge, i.e., an electric current, occurs. A conservation relationship holds between the electric charge and current, which will be described in Sect. 3.1.

1.2 Coulomb's Law

The electric force between electric charges is the **Coulomb force**. There are the following characteristic points between two electric charges in vacuum:

- The force between two electric charges of the same kind is repulsive and that between electric charges of different kinds is attractive.
- The magnitude of the force is proportional to the product of the two electric charges.
- The magnitude of the force is inversely proportional to the square of the distance between the two electric charges.
- The direction of the force lies on the straight line connecting the two electric charges.

This is called **Coulomb's law**: The force that electric charge q staying at position r experiences from the electric charge q' at position r' is:

$$F = \frac{qq'(r - r')}{4\pi\epsilon_0|r - r'|^3}, \tag{1.2}$$

where $(r - r')/|r - r'|$ is a unit vector directed from q' to q (see Fig. 1.1). The force that q' experiences from q is obtained by replacement of r by r' as $-F$. In the above the constant ϵ_0 is the **permittivity of vacuum**,

$$\epsilon_0 = \frac{10^7}{4\pi c^2} = 8.8541878 \times 10^{-12}\,\mathrm{C}^2/\mathrm{Nm}^2, \tag{1.3}$$

where c is the light speed in vacuum.

When there are more than two electric charges, the total Coulomb force is given by the sum of each individual Coulomb force. For example, the force on electric charge q at position r exerted by electric charges q_i at $r_i (i = 1, 2, \cdots, n)$ is

$$F = \sum_{i=1}^{n} F_i = \frac{q}{4\pi\epsilon_0} \sum_{i=1}^{n} \frac{q_i(r - r_i)}{|r - r_i|^3}, \tag{1.4}$$

Fig. 1.1 The Coulomb force exerted on electric charge q by electric charge q'

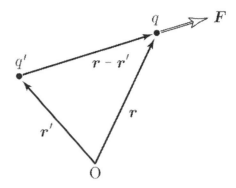

where $\boldsymbol{F}_i$ is the force exerted by the i-th electric charge (see Fig. 1.2). When the electric charge is distributed with the density $\rho(\boldsymbol{r}')$ within a region V in vacuum, we can regard the electric charge, $\rho dV'$, in an infinitesimal volume dV' as a point charge at position $\boldsymbol{r}'$ (see Fig. 1.3). Then, the Coulomb force that is exerted by the distributed electric charge on q outside the region V is:

$$\boldsymbol{F} = \frac{q}{4\pi\epsilon_0}\int_V \frac{\rho(\boldsymbol{r}')(\boldsymbol{r}-\boldsymbol{r}')}{|\boldsymbol{r}-\boldsymbol{r}'|^3}dV', \tag{1.5}$$

where $\int dV'$ is a volume integral with respect to $\boldsymbol{r}'$.

Fig. 1.2 The Coulomb force exerted on electric charge q by other electric charges

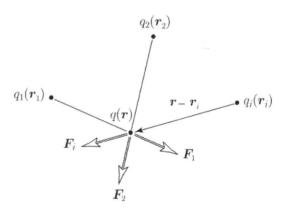

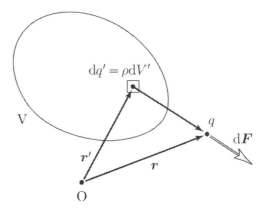

Fig. 1.3 The Coulomb force exerted on electric charge q by electric charge inside a small volume

1.3   Electric Field

When electric charge Q is placed in space, the Coulomb force is exerted on another charge q. It can be considered that this force is caused by an electrical distortion in space produced by the electric charge Q. Universal gravitation is also a nonlocal interaction similar to the Coulomb force. We are living under the Earth's gravity, and it can be regarded that this force is caused by a gravitational distortion produced by the large mass of the Earth. If we denote by G, M, and R_0 the constant of gravitation, the mass, and the radius of the Earth, the force that a particle of mass m on the surface of the earth receives is:

$$F_g = -G\frac{Mm i_r}{R_0^2} = -mg i_r, \tag{1.6}$$

where $g = 9.80665 \, \text{m/s}^2$ is the gravitational acceleration due to the Earth and $-i_r$ is a unit vector directed towards the center of the earth. The distorted property of space that gives rise to gravitation is called the gravitational field.

The electrical distortion caused by existing electric charges is called the **electric field**. When the force on the electric charge q placed at r due to the electric charge Q at the origin is expressed as

$$F = qE, \tag{1.7}$$

the vector E is called the **electric field strength**. The unit of the electric field is [N/C]. This also expressed as [V/m] using the unit [V] (**volt**) of electric potential, which will be defined later. Hence, the electric field strength is equal to the Coulomb force on a unit charge (1 C) that is placed at the aimed position. In the above case, we have

$$E(r) = \frac{Qr}{4\pi\epsilon_0|r|^3}. \tag{1.8}$$

When electric charges q_i are placed at $\mathbf{r}_i (i = 1, 2, \ldots, n)$, the electric field strength at $\mathbf{r}$ is

$$E(\mathbf{r}) = \frac{1}{4\pi \epsilon_0} \sum_{i=1}^{n} \frac{q_i (\mathbf{r} - \mathbf{r}_i)}{|\mathbf{r} - \mathbf{r}_i|^3}. \tag{1.9}$$

When the electric charge is distributed with density $\rho(\mathbf{r}')$ in area V, the electric field strength at $\mathbf{r}$ is

$$E(\mathbf{r}) = \frac{1}{4\pi \epsilon_0} \int_V \frac{\rho(\mathbf{r}')(\mathbf{r} - \mathbf{r}')}{|\mathbf{r} - \mathbf{r}'|^3} dV'. \tag{1.10}$$

Equations (1.9) and (1.10) for the electric field strength are also called Coulomb's law.

We can use **electric field lines** to visualize the electric field, which help us to understand the condition of the electric field. Such a line is defined so that the direction of a tangential line at any point on the electric field line is the same as that of the electric field E, and its line density is equal to the magnitude of E. Examples of electric field lines are shown in Fig. 1.4.

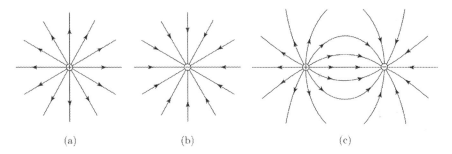

(a)　　　　　　　(b)　　　　　　　　　　　(c)

Fig. 1.4 Electric field lines of **a** a single positive charge, **b** a single negative charge, and **c** a pair of positive and negative charges of the same amount

---

Example 1.1 Two electric charges Q are placed at a distance $2a$ apart, as shown in Fig. 1.5. Determine the electric field strength at point A at the same distance from the two charges.

Fig. 1.5 Two electric
charges and observation
point A

Q•———————————•Q

Solution 1.1 The electric field strength exerted by one electric charge is $E' = Q/[4\pi\epsilon_0(a^2 + d^2)]$. Only the vertical component remains without being cancelled, as shown in the figure, and we have

$$E = 2E'\frac{d}{(a^2 + d^2)^{1/2}} = \frac{Qd}{2\pi\epsilon_0(a^2 + d^2)^{3/2}}.$$

◇

Example 1.2 Electric charge is uniformly distributed with density λ on a straight line, as shown in Fig. 1.6. Determine the electric field strength at point A at a distance a from the line.

Fig. 1.6 Uniformly
distributed electric charge on
a straight line and
observation point A

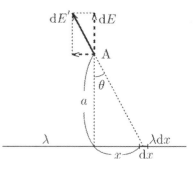

Solution 1.2 We define the x-axis on the line with the origin ($x = 0$) at the foot of the vertical line from point A. If we regard the electric charge in the region from x to $x + dx$, λdx, as a point charge, the electric field strength produced by this charge is

$$dE' = \frac{\lambda dx}{4\pi \epsilon_0 (x^2 + a^2)}.$$

The resultant electric field strength is determined as the sum of the contributions of all such electric charges. If we define angle θ as in the figure, the x-component of the electric field is cancelled due to symmetry, and only the vertical component, $dE = dE' \cos\theta$, remains. Using the relationship $x = a\tan\theta$, we have $dx = ad\theta/\cos^2\theta$ and $x^2 + a^2 = a^2/\cos^2\theta$. Thus, the electric field strength is determined to be

$$E = \int dE = \frac{\lambda}{2\pi \epsilon_0 a} \int_0^{\pi/2} \cos\theta d\theta = \frac{\lambda}{2\pi \epsilon_0 a} [\sin\theta]_0^{\pi/2} = \frac{\lambda}{2\pi \epsilon_0 a}. \tag{1.11}$$

$\diamond$

1.4 Electric Potential

We suppose that electric charge q is forced to move from point A at r_A to point B at r_B in the electric field E. It is necessary to apply an opposite force, $-qE$, to the electric charge to prevent it from moving by the Coulomb force. So, the work necessary to carry the electric charge is

$$W = -q \int_{r_A}^{r_B} E \cdot dr. \tag{1.12}$$

Here, we treat the case where the electric field is given by Eq. (1.8) for point charge Q at the origin, as shown in Fig. 1.7. In this case, using the angle θ and $d\hat{r}$, Eq. (1.12) leads to

Fig. 1.7 Movement of an electric charge in an electric field

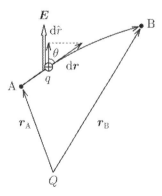

$$W = -\frac{qQ}{4\pi\epsilon_0} \int_{r_A}^{r_B} \frac{\cos\theta \, dr}{r^2} = -\frac{qQ}{4\pi\epsilon_0} \int_{r_A}^{r_B} \frac{d\hat{r}}{r^2} = \frac{qQ}{4\pi\epsilon_0} \left(\frac{1}{r_B} - \frac{1}{r_A} \right), \qquad (1.13)$$

where $r_A = |\mathbf{r}_A|$ and $r_B = |\mathbf{r}_B|$. Hence, the needed work is determined only by the starting point A and the terminal point B but is independent of a route of integration. This result holds generally, also for the case where many electric charges are distributed in space or the electric charge is distributed continuously. Here, we introduce a quantity defined by

$$\phi(\mathbf{r}) = -\int_{r_0}^{r} \mathbf{E} \cdot d\mathbf{r}, \qquad (1.14)$$

which is called the **electric potential**. r_0 is a reference point at which the electric potential is zero and is usually taken as at infinity. The unit of the electric potential is [V]. The value of the electric potential is determined by the starting point and terminal point only and is independent of the route. Hence, the electric potential at an arbitrary point is equal to the work necessary to carry a unit charge from infinity. Thus, the work given by Eq. (1.12) is generally expressed using the electric potential as

$$W = q[\phi(\mathbf{r}_B) - \phi(\mathbf{r}_A)]. \qquad (1.15)$$

The difference in the electric potential between two points, such as $\phi(\mathbf{r}_B) - \phi(\mathbf{r}_A)$, is called the **voltage**. If ϕ does not change while carrying q from infinity to $\mathbf{r}$, the work is written as

$$W = q\phi(\mathbf{r}). \qquad (1.16)$$

This can be regarded as the potential energy of electric charge q in the electric potential ϕ. This is called the **electrostatic energy**.

Using Eq. (1.14), we have

$$\oint_C \mathbf{E} \cdot d\mathbf{r} = 0 \qquad (1.17)$$

for an arbitrary closed loop C, where the integral shows a circular integral. The field with such a property is called a conservative field. This means that the work needed to carry an electric charge along a closed loop is zero, which is equivalent to the zero mechanical work needed to carry a mass along a closed loop in the gravitational field.

Here, we determine the electric potential in space when electric charge Q is placed at the origin. The electric field is given by Eq. (1.8). Using Eq. (1.14), we have

$$\phi(\boldsymbol{r}) = -\int_{\infty}^{r} \frac{Q}{4\pi\epsilon_0 r'^2} dr' = \frac{Q}{4\pi\epsilon_0 r}. \tag{1.18}$$

When there are many electric charges and the electric field is given by Eq. (1.9), the electric potential leads to

$$\phi(\boldsymbol{r}) = \frac{1}{4\pi\epsilon_0} \sum_{i=1}^{n} \frac{q_i}{|\boldsymbol{r} - \boldsymbol{r}_i|}. \tag{1.19}$$

When the electric charge is continuously distributed with density ρ, the electric field obeys Eq. (1.10) and the electric potential is

$$\phi(\boldsymbol{r}) = \frac{1}{4\pi\epsilon_0} \int_V \frac{\rho(\boldsymbol{r}')}{|\boldsymbol{r} - \boldsymbol{r}'|} dV'. \tag{1.20}$$

When the electric potential is given, the electric field can be determined. The details are described in Appendix A.1. Using Cartesian coordinates, the electric field is described as

$$\boldsymbol{E} = -\frac{\partial\phi}{\partial x}\boldsymbol{i}_x - \frac{\partial\phi}{\partial y}\boldsymbol{i}_y - \frac{\partial\phi}{\partial z}\boldsymbol{i}_z, \tag{1.21}$$

where $\boldsymbol{i}_x$, $\boldsymbol{i}_y$, and $\boldsymbol{i}_z$ are unit vectors along the x, y, and z-axes.

A virtual surface on which the electric potential has the same value is called the **equipotential surface**. Examples of equipotential surface are shown in Fig. 1.8. An equipotential surface is perpendicular to the electric field lines. This can be proved as follows: The work necessary to carry electric charge q by a small distance $\Delta\boldsymbol{r}$

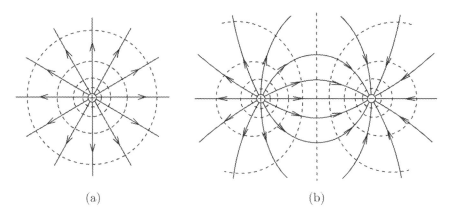

(a)   (b)

Fig. 1.8 The dotted lines show examples of equipotential surfaces on a plane that includes electric charges. The solid lines are electric field lines

in electric field E is $-qE \cdot \Delta r$. When Δr is on an equipotential surface, this work must be zero from Eq. (1.15). Thus, we can show that Δr is perpendicular to E, and hence, to the electric field lines.

Example 1.3 When two electric charges Q are placed at a distance $2a$ apart, as shown in Fig. 1.5 in Example 1.1, determine the electric potential at point A at the same distance from the two charges.

Solution 1.3 We have two solution methods. First, using Solution 1.1, the electric field at a point at distance z from the center of the two electric charges is directed along the z-axis and its strength is

$$E = \frac{Qz}{2\pi \epsilon_0 (a^2 + z^2)^{3/2}}.$$

Hence, using Eq. (1.14), we have

$$\phi = -\int_\infty^d \frac{Qz}{2\pi \epsilon_0 (a^2 + z^2)^{3/2}} dz = \left[\frac{Q}{2\pi \epsilon_0 (a^2 + z^2)^{1/2}} \right]_\infty^d = \frac{Q}{2\pi \epsilon_0 (a^2 + d^2)^{1/2}}.$$

Second, we can directly determine the electric potential using Eq. (1.19). The electric potential produced by one electric charge is $\phi' = Q / \left[2\pi \epsilon_0 (a^2 + d^2)^{1/2} \right]$. Since there is another electric charge at the same distance, and we have the same result:

$$\phi = 2\phi' = \frac{Q}{2\pi \epsilon_0 (a^2 + d^2)^{1/2}}.$$

◇

Example 1.4 When three electric charges Q are placed at each vertex of an equilateral triangle with side length a, as shown in Fig. 1.9, determine the electric potential difference between point A at the center and point B on the midpoint of one side.

Fig. 1.9 Electric charges placed at vertices of an equilateral triangle

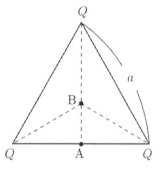

Solution 1.4 Point A is placed at distance $a/2$ from two electric charges and $(\sqrt{3}/2)a$ from the other electric charge. Hence, the electric potential at point A is

$$\phi_A = \frac{Q}{2\pi\epsilon_0 a} \times 2 + \frac{Q}{2\sqrt{3}\pi\epsilon_0 a} = \frac{\left(6+\sqrt{3}\right)Q}{6\pi\epsilon_0 a}.$$

Point B is placed at distance $a/\sqrt{3}$ from all the electric charges, and the electric potential there is

$$\phi_B = \frac{\sqrt{3}Q}{4\pi\epsilon_0 a} \times 3 = \frac{3\sqrt{3}Q}{4\pi\epsilon_0 a}.$$

Thus, we have

$$\phi_A - \phi_B = -\frac{\left(7\sqrt{3}-12\right)Q}{12\pi\epsilon_0 a}.$$

Note that $\phi_A - \phi_B < 0$.

$\diamond$

1.5   Gauss's Law

We suppose a closed surface S that includes charge q inside, and try to determine the surface integral on S of $\boldsymbol{E}$:

$$N = \int_S \boldsymbol{E} \cdot \mathrm{d}\boldsymbol{S} = \int_S E_n \mathrm{d}S, \tag{1.22}$$

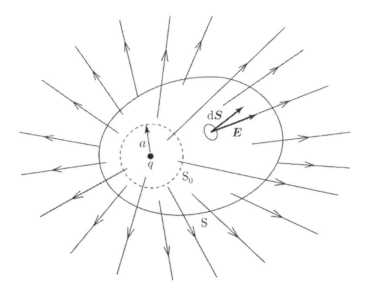

Fig. 1.10 Electric field lines going out of S, which includes electric charge q. S_0 is a sphere of radius a with its origin at the position of the electric charge

where E_n is the normal component of the electric field directed outward on S (see Fig. 1.10).

N in Eq. (1.22) is called the number of electric field lines that go out of S. Since the electric field lines do not terminate or appear halfway, the surface integral is the same as if it is calculated on a sphere S_0 of radius a with its center at the position of the electric charge q. The electric field E is parallel to the elementary surface dS and takes on a constant value, $q/\left(4\pi\epsilon_0 a^2\right)$, on S_0, so a simple calculation leads to

$$N = \frac{q}{4\pi\epsilon_0 a^2}S = \frac{q}{\epsilon_0}, \tag{1.23}$$

where $S = 4\pi a^2$ is the surface area of S_0.

If the closed surface S does not include the electric charge q, as shown in Fig. 1.11, the electric field lines that go into S surely go out, and hence, N is zero.

Extending the above results, when electric charges $q_1, q_2, \ldots, q_m$ are inside S and $q_{m+1}, q_{m+2}, \ldots, q_{m+n}$ are outside S, the number of electric field lines that go out of S is

$$N = \int_S E \cdot dS = \frac{1}{\epsilon_0}\sum_{i=1}^{m} q_i, \tag{1.24}$$

and there is no contribution from electric charges $q_{m+1}, q_{m+2}, \ldots, q_{m+n}$. This is given by the total electric charge inside S divided by ϵ_0. When the electric charge is

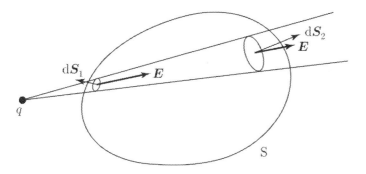

Fig. 1.11 Electric field lines going in and out of a closed surface S that does not include electric charge q

distributed with density $\rho(\boldsymbol{r})$, we have

$$N = \int_S \boldsymbol{E} \cdot \mathrm{d}\boldsymbol{S} = \frac{1}{\epsilon_0} \int_V \rho(\boldsymbol{r})\mathrm{d}V, \tag{1.25}$$

where V is the volume surrounded by S. Equations (1.24) and (1.25) are called **Gauss's law**.

Here, we show an example of how to determine the electric field strength using Gauss's law. Suppose that electric charge is uniformly distributed with density ρ inside a sphere of radius a. Although it is not easy to determine the electric field strength due to this electric charge with Coulomb's law, we can easily determine it with Gauss's law. In the beginning, we shall determine the electric field strength at a point at distance $r\,(< a)$ from the center of the sphere. We assume a concentric sphere S of radius r, as shown in Fig. 1.12a. Then, the observation point is placed on S. If we apply Gauss's law on S, the electric field vector has only the radial component, and its value E is constant on S from symmetry. Hence, the left side of Eq. (1.25) leads to

$$\int_S E\mathrm{d}S = 4\pi r^2 E. \tag{1.26}$$

On the other hand, the total electric charge inside S is $(4/3)\pi r^3 \rho$. Hence, the electric field strength is determined to be

$$E = \frac{\rho r}{3\epsilon_0}. \tag{1.27}$$

Next, we shall determine the electric field strength outside the sphere. We assume a concentric sphere with radius $r\,(> a)$, as in Fig. 1.12b and apply Gauss's law on this spherical surface. In this case, the left side of Eq. (1.25) is the same as Eq. (1.26),

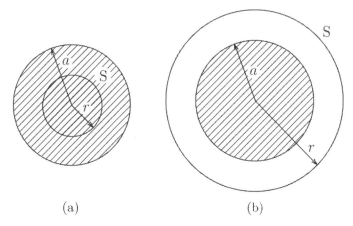

(a)   (b)

Fig. 1.12 Method used to apply Gauss's law to determine the electric field strength when the electric charge is uniformly distributed with density ρ inside a sphere of radius a for the cases of **a** inside the sphere and **b** outside the sphere

and the total electric charge inside S is $(4/3)\pi a^3 \rho$. Thus, the electric field strength is given by.

$$E = \frac{\rho a^3}{3\epsilon_0 r^2}. \tag{1.28}$$

It is found that Eq. (1.28) is identical to Eq. (1.8) if we use the total electric charge, $Q = (4/3)\pi a^3 \rho$. That is, the result is same as in the case in which all the electric charge is concentrated at the origin. This corresponds to the fact that the gravity on a matter on the Earth is the same as the gravity when all the mass is concentrated on the center of the Earth, as shown in Eq. (1.6).

The electric potential is discussed here. Since the electric field is directed radially, the electric potential is given by

$$\phi(r) = -\int_{\infty}^{r} E(r)\mathrm{d}R. \tag{1.29}$$

Simple calculation yields

$$\phi(r) = -\int_{\infty}^{r} \frac{\rho a^3}{3\epsilon_0 r^2}\mathrm{d}r = \frac{\rho a^3}{3\epsilon_0 r}; \qquad\qquad r > a,$$

$$= -\int_{\infty}^{r} \frac{\rho r}{3\epsilon_0}\mathrm{d}r + \frac{\rho a^2}{3\epsilon_0} = \frac{\rho}{2\epsilon_0}\left(a^2 - \frac{r^2}{3}\right); \qquad 0 \le r < a. \tag{1.30}$$

Distributions of the electric field strength and electric potential are shown in Fig. 1.13a, b.

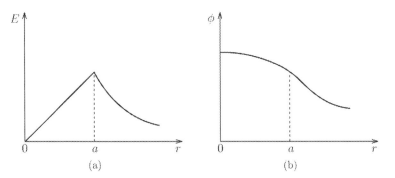

(a)   (b)

Fig. 1.13 Distributions of **a** electric field strength and **b** electric potential when electric charge is uniformly distributed inside a sphere of radius a

---

Example 1.5 Suppose that electric charge is uniformly distributed with surface density σ on a wide plane. Determine the electric field strength at point A at a distance a from the plane.

Solution 1.5 Assume a closed cylindrical surface, S, of radius a and height $2h$, which includes the plane, as shown in Fig. 1.14. Point A is placed on the top surface of S. We apply Gauss's law on this surface. From symmetry, the electric field lines are directed normal to the plane, and there is no electric field line that goes out of the side surface. So, all the electric field lines go out of the top and bottom surfaces. The electric field strength on the surfaces is denoted by E. Then, the surface integral in Eq. (1.25) is $2\pi a^2 E$. Since the total electric charge inside S is $\pi a^2 \sigma$, Eq. (1.25) leads to

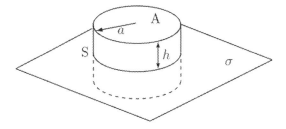

Fig. 1.14 Plane with uniformly distributed electric charge and cylindrical surface

$$E = \frac{\sigma}{2\epsilon_0}.$$

Thus, the electric field strength does not depend on the distance from the plane.

◇

Example 1.6 Solve the problem of Example 1.2 with Gauss's law.

Solution 1.6 Assume a closed cylindrical surface S of radius a and length l with the central axis on the linear electric charge, as shown in Fig. 1.15. Gauss's law is applied on S. The electric field lines are directed radially from the linear charge and takes on the same value on the cylindrical side surface S'. The electric field is normal to dS on the two edge surfaces, and there is no contribution from these surfaces to the surface integral. Thus, the surface integral reduces to

$$\int_S \boldsymbol{E} \cdot \mathrm{d}\boldsymbol{S} = \int_{S'} E\,\mathrm{d}S = 2\pi a l E.$$

The total electric charge inside S is λl. Thus, from Gauss's law we have

$$E = \frac{\lambda}{2\pi \epsilon_0 a}.$$

Fig. 1.15 Closed cylindrical surface with the central axis on the linear electric charge

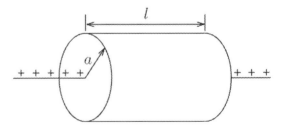

◇

Exercises

1.1 Electric charges $\pm Q$ are placed at a distance $2a$, as shown in Fig. 1.16. Determine the electric field strength and electric potential at point A equidistant from the two electric charges.

Fig. 1.16 Two electric
charges and observation
point A

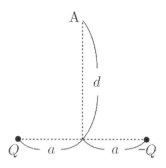

1.2 Three electric charges Q and three electric charges $-Q$ are placed at points
   on the three axes at equidistance a from the origin O, as shown in Fig. 1.17.
   Determine the electric field strength and electric potential at the origin O.

1.3 Electric charge Q and unknown electric charges Q_x and Q_y are placed at each
   vertex of an equilateral triangle with side length a, as shown in Fig. 1.18.
   The electric field strength at point A, which is symmetric with the position of
   the charge Q_x, is zero. Determine the values of Q_x and Q_y and the electric
   potential at point A.

Fig. 1.17 Arrangement of 6
electric charges

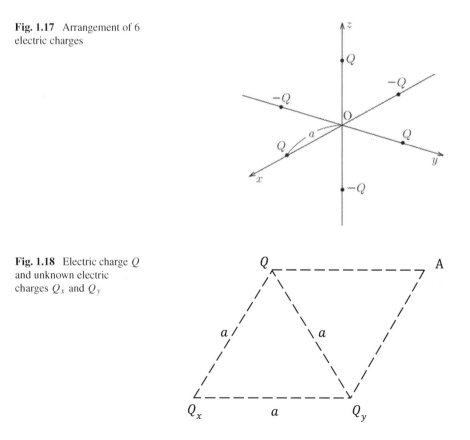

Fig. 1.18 Electric charge Q
and unknown electric
charges Q_x and Q_y

Fig. 1.19 Circle with
uniform linear electric
charge

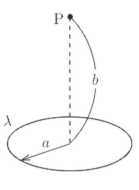

Fig. 1.20 Electric charge of
density σ on a disk of radius
R

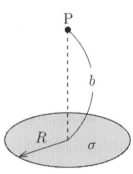

1.4 Electric charge is uniformly distributed with linear density λ on a circle of
radius a, as shown in Fig. 1.19. Determine the electric field strength and
electric potential at point P at a distance b from the center of the circle.

1.5 Electric charge is uniformly distributed with surface density σ on a disk of
radius R, as shown in Fig. 1.20. Determine the electric field strength and
electric potential at point P at a distance b from the center of the disk.

1.6 Solve the problem of Example 1.5 using Coulomb's law.

Fig. 1.21 Uniformly
distributed electric charge
with linear density λ on a
square of side length a and
observation point P

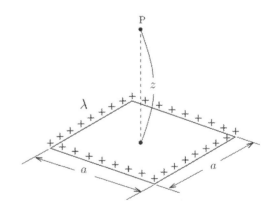

Fig. 1.22 Electric charge distributed uniformly with density σ on a long plate of width $2w$

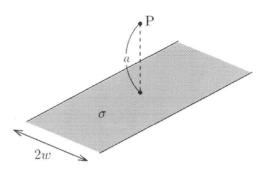

1.7 Electric charge is uniformly distributed with linear density λ on a square of side length a, as shown in Fig. 1.21. Determine the electric field strength at point P at a distance z from the center.

1.8 Electric charge is uniformly distributed with surface density σ on a long plate of width $2w$, as shown in Fig. 1.22. Determine the electric field strength at point P at a distance a from the center of the plate. (*Hint*: Divide the planar electric charge into a set of thin line charges and apply Gauss's law to each line charge.)

1.9 When electric charge is uniformly distributed with surfaced density σ on a sphere of radius a, determine the electric potential at the center of the sphere.

1.10 Electric charge is uniformly distributed with density ρ in a hollow cylinder ($a \leq R \leq b$), as shown in Fig. 1.23. Determine the electric field and electric potential in each area. The reference point at which the electric potential is zero is taken as at $R = R_\infty (\gg b)$. The reason why the reference point is not taken as at infinity is to avoid a divergence of the electric potential due to existence of infinite amount of electric charge in space.

1.11 Electric charge is uniformly distributed with density ρ in a solid portion of the sphere of radius a with a spherical void of radius b, as shown in Fig. 1.24. Determine the electric field strength and electric potential at point P, which is placed at a distance d from the center of the sphere on the line connecting the center and the void center. (*Hint*: The result is obtained by a superposition of the electric charge with density ρ in the solid sphere and that with density $-\rho$ in the void.)

Fig. 1.23 Cross-section of a long hollow cylinder in which electric charge is uniformly distributed with density ρ

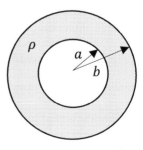

Fig. 1.24 Electric charge
distributed uniformly in a
solid portion of a sphere of
radius b with a spherical void
and observation point P

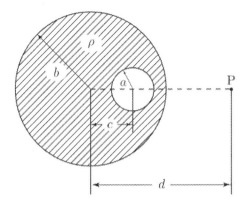

Chapter 2
Conductors and Dielectric Materials

Abstract This chapter covers electric phenomena in electric materials, which are classified into conductors and dielectric materials. When an external electric field is applied to a conductor, the electric field strength is zero inside the conductor due to induced electric charges on the surface. Conductors are equipotential. When an external electric field is applied to a dielectric material, the electric field penetrates the dielectric material, although its strength is weakened by polarization charges, which cannot move freely due to bonding to nuclei. The refraction of electric field lines at an interface between different dielectric materials is explained. Capacitors made of such dielectric materials are introduced and we learn the capacitance of capacitors, which is the electric charge that can be stored by a unit electric potential difference, 1 V.

2.1 Conductors

Electric materials are classified into **conductors** that can easily transport current and **insulators** that can hardly do so. Representative conductor materials are metals. Their electric properties originate from free electrons, i.e., true charges, that can move freely in the material. On the other hand, electrons in insulators cannot move freely because of their bonding to atomic nuclei. Hence, the electric behavior is quite different between conductors and insulators. Electrons in a conductor can move due to the Coulomb force in the electric field and are distributed on the conductor surface in a steady state. In this case the applied electric field and the electric field produced by the distributed electric charge cancel each other, resulting in zero electric field inside the conductor. If a part of the electric field remains, it surely drives true charges, resulting in a contradiction of the hypothesis of a steady state. If some true charges remain in the interior of the conductor, they produce an electric field, resulting in a similar contradiction. Such an induction of electric charges on the conductor surface under an applied electric field is called **electrostatic induction**.

Fig. 2.1 Condition of
electric field near the surface
of a conductor

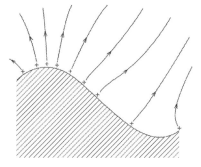

In an insulator in an applied electric field, electric charges are slightly displaced by the Coulomb force but cannot move freely. Thus, the electric field cannot be shielded completely and penetrates the material. Insulators are also called **dielectrics** or **dielectric materials**.

As shown above, the electric field strength and electric charge density are zero inside a conductor:

$$\boldsymbol{E} = 0, \tag{2.1}$$

$$\rho = 0. \tag{2.2}$$

Equation (2.1) leads to

$$\phi = c(\text{const.}). \tag{2.3}$$

That is, the conductor is **equipotential**. So, when an electric field exists around a conductor, is must be normal to its surface, as illustrated in Fig. 2.1.

Here, we assume the case in which electric field E is applied normal to a conductor in the form of a wide slab. Before application, positive and negative electric charges of equal density are distributed uniformly, and an electrically neutral state was achieved. Positive and negative electric charges are slightly displaced in opposite directions by the applied electric field, as shown in Fig. 2.2a. Thus, the electrically neutral state is maintained in the region where the distributions of positive and negative electric charges are overlapping, and electric charges appear on the surfaces of the conductor, as shown in Fig. 2.2b. We denote by $\pm\sigma$ the densities of electric charges that appear on the surfaces of the conductor. Here, we determine the relation between σ and the applied electric field strength E. We apply Gauss's law on a closed surface, as shown in Fig. 2.3. The electric field strength is E outside the conductor and 0 inside the conductor. Since the electric field lines do not go out of the side surface, the surface integral of the electric field strength is equal to ES, where S is the area of the conductor surface inside the closed surface. Since the total electric charge inside the closed surface is σS, we have

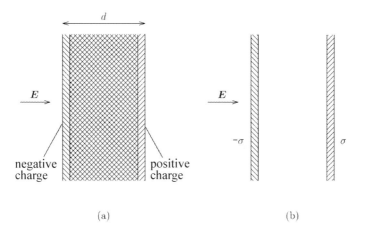

Fig. 2.2 **a** Movement of electric charges inside a wide slab conductor in an applied normal electric field and **b** electric charges on the surfaces

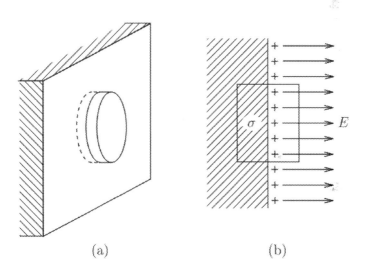

Fig. 2.3 **a** Closed surface on which Gauss's law is applied and **b** electric field near the conductor surface

$$\sigma = \epsilon_0 E. \tag{2.4}$$

The electric charges with densities $\pm\sigma$ on the surface cancel the applied electric field inside the conductor.

Here, we consider the case in which an electric charge Q is given to a spherical conductor of radius a. Electric charge is uniformly distributed on the surface of the conductor, so the electric field does not appear inside the conductor. From symmetry, the electric field is directed radially outward, and we can determine the electric field strength using Gauss's law, as was done in Sect. 1.5. This yields

$$E(r) = \frac{Q}{4\pi \epsilon_0 r^2}; \quad r > a,$$
$$= 0; \qquad 0 \le r < a. \tag{2.5}$$

The electric potential is determined to be

$$\phi(r) = \frac{Q}{4\pi \epsilon_0 r}; \quad r > a,$$
$$= \frac{Q}{4\pi \epsilon_0 a}; \quad 0 \le r < a. \tag{2.6}$$

The obtained results for the electric field strength and electric potential are shown in Figs. 2.4 and 2.5, respectively.

Next, we treat the case in which electric charge Q is given to the inner conductor of concentric spherical conductors, as shown in Fig. 2.6. The electric charge is uniformly distributed on the surface of the inner conductor, as discussed in the above example.

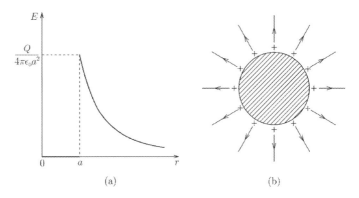

(a)   (b)

Fig. 2.4 a Distribution of electric field strength and **b** electric field lines when electric charge is given to a spherical conductor

Fig. 2.5 Distribution of electric potential when electric charge is given to a spherical conductor

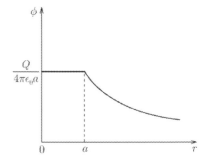

Fig. 2.6 Concentric
spherical conductors

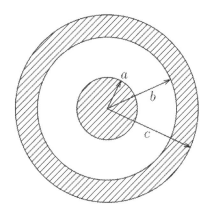

Then, electric charge appears on the inner surface of the outer conductor ($r = b$) so
that the electric field does not penetrate the outer conductor. This electric charge is
denoted by Q_b. Here, we assume a concentric virtual sphere of radius r ($b < r < c$)
and apply Gauss's law on it. Since the virtual sphere is inside the outer conductor, the
electric field strength is zero on it. Thus, the total electric charge inside the sphere,
$Q + Q_b$, is zero, and we have $Q_b = -Q$. Since no electric charge is given to the
outer conductor, the electric charge that appears on the outer surface $r = c$ is Q.
Thus, electric charges $-Q$ and Q are uniformly distributed on the surfaces of $r = b$
and $r = c$, respectively. Then, the electric field strength and electric potential are

$$
\begin{aligned}
E(r) &= \frac{Q}{4\pi\epsilon_0 r^2}; & r > c, \\
&= 0; & b < r < c, \\
&= \frac{Q}{4\pi\epsilon_0 r^2}; & a < r < b, \\
&= 0; & 0 \le r < a,
\end{aligned}
\tag{2.7}
$$

and

$$
\begin{aligned}
\phi(r) &= \frac{Q}{4\pi\epsilon_0 r}; & r > c, \\
&= \frac{Q}{4\pi\epsilon_0 c}; & b < r < c, \\
&= \frac{Q}{4\pi\epsilon_0}\left(\frac{1}{r} - \frac{1}{b} + \frac{1}{c}\right); & a < r < b, \\
&= \frac{Q}{4\pi\epsilon_0}\left(\frac{1}{a} - \frac{1}{b} + \frac{1}{c}\right); & 0 \le r < a.
\end{aligned}
\tag{2.8}
$$

The obtained electric field strength and electric potential are shown in Fig. 2.7.

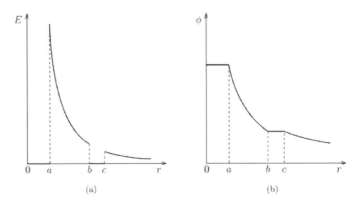

Fig. 2.7 **a** Electric field strength and **b** electric potential when electric charge is given to the inner conductor of concentric spherical conductors

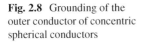

Fig. 2.8 Grounding of the outer conductor of concentric spherical conductors

Here, we considered the case in which the outer conductor is grounded (see Fig. 2.8). Grounding is a method to make the electric potential of a conductor zero by connecting it to the ground. Since the ground is sufficiently large, electric charge that can move is transferred to the ground. In the present case, the electric charge at $r = c$ is transferred to the ground. The electric charge at $r = b$ does not transfer, since it shields penetration of the electric field into the outer conductor. The electric field strength and electric potential in this case are

$$E(r) = 0; \qquad r > b,$$
$$= \frac{Q}{4\pi\epsilon_0 r^2}; \quad a < r < b,$$
$$= 0; \qquad 0 \leq r < a, \tag{2.9}$$

and

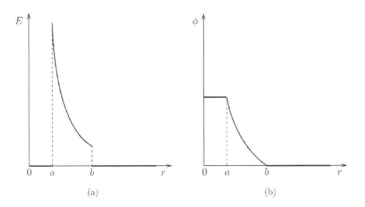

Fig. 2.9 a Electric field strength and **b** electric potential when electric charge is given to the inner conductor of concentric spherical conductors and the outer conductor is grounded

$$\phi(r) = 0; \qquad\qquad\qquad r > b,$$
$$= \frac{Q}{4\pi\epsilon_0}\left(\frac{1}{r} - \frac{1}{b}\right); \quad a < r < b,$$
$$= \frac{Q}{4\pi\epsilon_0}\left(\frac{1}{a} - \frac{1}{b}\right); \quad 0 \leq r < a. \qquad (2.10)$$

These results are shown in Fig. 2.9.

The latter example shows that, when electric charge is surrounded by a grounded conductor, the effect of the inner electric charge does not appear outside. Or if some area is completely surrounded by a grounded conductor, the electric effect from outside does not go into the area. This is called **electrostatic shielding**.

Example 2.1 Suppose that electric charges Q_1 and Q_2 are given to the inner and outer conductors of the concentric spherical conductors in Fig. 2.6. Determine the electric field strength and electric potential in each region.

Solution 2.1 First, we determine the distribution of electric charges. Electric charge Q_1 is uniformly distributed on the surface of the inner conductor ($r = a$). The electric charge induced on the inner surface of the outer conductor at $r = b$ is determined to be $-Q_1$ using Gauss's law, and the principle of conservation of electric charge implies that the electric charge on the outer surface at $r = c$ is $Q_1 + Q_2$. Thus, the electric field strength is obtained to be

$$E(r) = \frac{Q_1 + Q_2}{4\pi\epsilon_0 r^2}; \qquad r > c,$$
$$= 0; \qquad\qquad b < r < c,$$

$$= \frac{Q_1}{4\pi \epsilon_0 r^2}; \qquad a < r < b,$$

$$= 0; \qquad 0 \le r < a.$$

The electric potential is

$$\phi(r) = \frac{Q_1 + Q_2}{4\pi \epsilon_0 r}; \qquad\qquad r > c,$$

$$= \frac{Q_1 + Q_2}{4\pi \epsilon_0 c}; \qquad\qquad b < r < c,$$

$$= \frac{Q_1}{4\pi \epsilon_0}\left(\frac{1}{r} - \frac{1}{b} + \frac{1}{c}\right) + \frac{Q_2}{4\pi \epsilon_0 c}; \quad a < r < b,$$

$$= \frac{Q_1}{4\pi \epsilon_0}\left(\frac{1}{a} - \frac{1}{b} + \frac{1}{c}\right) + \frac{Q_2}{4\pi \epsilon_0 c}; \quad 0 \le r < a.$$

◇

Example 2.2 We give electric charge Q to each wide slab conductor in Fig. 2.10. Determine the electric charge on each surface and electric field strength in each region. The area of each surface is S.

Fig. 2.10 Two wide slab conductors

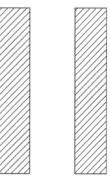

Solution 2.2 We denote by Q_x the electric charge that appears on the right surface of the left slab conductor. Then, the electric charge on the left surface of this conductor is $Q - Q_x$ and the total electric charge that exist on the right side of this conductor is $Q + Q_x$. The two electric charges must be the same so that no electric field appears in this conductor. Thus, we have

$$Q_x = 0.$$

So, the electric charge on the left surface of this conductor is Q. The electric charge distribution in the right conductor can also be obtained in a similar manner, and the electric charges on each surface are Q, 0, 0, and Q from the left. Thus, the electric field strength directed to the right in each region is

$$E = -\frac{Q}{\epsilon_0 S}; \quad \text{left vacuum,}$$

$$= 0; \quad \text{two conductors and space inbetween,}$$

$$= \frac{Q}{\epsilon_0 S}; \quad \text{right vacuum.}$$

◇

2.2 Image Method

It is supposed that an electric charge q is placed at a distance a from an infinitely wide conductor surface, as shown in Fig. 2.11a. The opposite electric charge is induced on the conductor surface to shield the inside against the electric field produced by the applied electric charge. It may seem to be difficult to determine the electric charge distribution on the conductor surface and the electric field strength in the space. There is a special effective method to solve this problem, however. This is the **image method**, in which we use the equipotential property on the conductor surface and Eq. (2.4).

Here, we use Cartesian coordinates with the x-y plane ($z = 0$) on the conductor surface and the origin, O, at the foot of a perpendicular line from the electric charge. If there were no conductor, we would only have to place the opposite electric charge $-q$ at the position symmetric with respect to the conductor surface, as shown in Fig. 2.11b, to satisfy the condition of equipotential at the position of the conductor surface. Such a method of virtual removal of the conductor accompanied by placing an opposite electric charge at a mirror image position with respect to the conductor

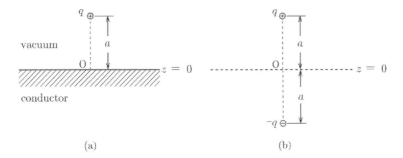

(a)   (b)

Fig. 2.11 **a** Electric charge q placed at a distance from wide conductor surface and **b** image charge $-q$ placed at a mirror position with respect to the conductor surface after virtually removing the conductor

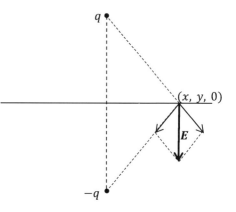

Fig. 2.12 Electric field on the conductor surface produced by the given electric charge and the image charge

surface is the image method. This method is based on the fact that a plane with the same distance from positive and negative electric charges of the same magnitude, which is now the position of the conductor surface, is equipotential, as shown in Fig. 1.8b. The virtual electric charge is called the **image charge**. The electric field strength and electric potential obtained by this method are correct only in vacuum ($z > 0$). The electric field strength on the conductor surface produced by the given electric charge and image charge is normal to the surface, as shown in Fig. 2.12, and its strength is

$$E_z(x, y, 0) = -\frac{qa}{2\pi \epsilon_0 (x^2 + y^2 + a^2)^{3/2}}. \tag{2.11}$$

The density of electric charge on the conductor surface is obtained from Eq. (2.3) as

$$\sigma = \epsilon_0 E_z(x, y, 0) = -\frac{qa}{2\pi (x^2 + y^2 + a^2)^{3/2}}. \tag{2.12}$$

The electric field strength and electric potential in vacuum can be obtained using the given electric charge and image charge (see Exercise 2.4).

Here, we shall determine the total electric charge induced on the conductor surface. We use two-dimensional polar coordinates expressed as $r^2 = x^2 + y^2$. Then, the total electric charge on the surface is

$$q' = \int dx \int dy \sigma = -\frac{qa}{2\pi} \int_0^\infty \frac{2\pi r \, dr}{(r^2 + a^2)^{3/2}} = -qa\left[-\frac{1}{(r^2 + a^2)^{1/2}} \right]_0^\infty = -q. \tag{2.13}$$

So, it is found that the total electric charge is equal to the image charge.

Next, we shall calculate the Coulomb force between the given charge and the electric charge induced on the conductor surface. First, we determine the force due to the electric charge distributed on the thin circle in the region from r to $r + dr$ from the origin. From symmetry, only the vertical component remains for the force due to the electric charge of the azimuthal angle from φ to $\varphi + d\varphi$, $dQ = \sigma(r)r\,dr\,d\varphi$. Thus, the Coulomb force due to the electric charge on the thin circle is given by

$$dF = \frac{q}{4\pi\epsilon_0} \int_0^{2\pi} \frac{a\sigma(r)r\,dr}{(r^2 + a^2)^{3/2}}\,d\varphi = -\frac{q^2 a^2}{4\pi\epsilon_0} \cdot \frac{r\,dr}{(r^2 + a^2)^3}. \qquad (2.14)$$

Integrating this with respect to r from 0 to infinity, we have the Coulomb force:

$$F = -\frac{q^2 a^2}{4\pi\epsilon_0} \int_0^\infty \frac{r\,dr}{(r^2 + a^2)^3} = -\frac{q^2}{16\pi\epsilon_0 a^2}. \qquad (2.15)$$

This is equal to the Coulomb force between the given charge and the image charge.

Example 2.3 The electric charge on the conductor surface was determined above using the image method. Prove that the interior of the conductor is completely shielded by the induced electric charge given by Eq. (2.12).

Solution 2.3 The electric charge induced on the conductor surface also produces the electric field inside the conductor. This electric field is symmetric to that in vacuum with respect to the surface. Hence, the electric field in the conductor produced by the induced electric charge is equal to that produced by an electric charge $-q$ placed at the position of q. Namely, the total electric field strength is equal to that produced by q and $-q$ at the same place, i.e., the electric field strength when no electric charge is given. Thus, the zero electric field strength in the conductor can be proved.

$\diamond$

2.3 Capacitor and Capacitance

Elements used to store electric charge in electric circuits are called **capacitors** or **condensers**. A capacitor is composed of two conductors. Since electric charge stays only on conductor surfaces, usually conductors have plate shapes with wide area. In addition, the distance between two plate conductors is kept small in order to store electric charges effectively with a small voltage, as shown in Fig. 2.13. A capacitor of such a type is called a **parallel-plate capacitor**. Usually, the distance between the two plate conductors, which are called electrodes, d, is much smaller the size of an electrode, $S^{1/2}$, where S is its surface area.

Fig. 2.13 Structure of
parallel-plate capacitor

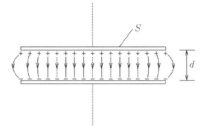

Suppose that we give electric charge Q to one electrode of a capacitor, and ground the other electrode. Then, electric charge moves to the other electrode from the ground due to electrostatic induction. The electric charges are uniformly distributed facing each other on the two electrodes, as shown in Fig. 2.14. The density of electric charge on the upper electrode is Q/S, and if we apply Gauss's law on a closed surface (a) shown by dotted line, it can be easily derived that the electric charge density on the lower electrode is equal to $-Q/S$. That is, since the upper and lower surfaces are inside the conductor and the electric field is parallel to the side surface of the closed surface, the electric field lines do not go outside. This shows that the total electric charge is zero inside the closed surface. Thus, the electric field is limited to within the narrow region between the two electrodes. If we add electric charge $-Q$ to the lower electrode instead of grounding it, we have the same condition.

Here, we shall determine the **capacitance**, i.e., the electric charge that can be stored in a capacitor with a unit electric potential difference. Assume again that we give electric charge Q to one electrode and that the other electrode is grounded, as shown in Fig. 2.14. When we apply Gauss's law on the closed surface (b), the electric field strength in the space between the electrodes is determined to be

$$E = \frac{Q}{\epsilon_0 S}. \tag{2.16}$$

This result can also be derived from Eq. (2.4). Thus, the electric potential difference between the two electrodes is

Fig. 2.14 Distribution of
electric charge in the two
electrodes and the electric
field between them

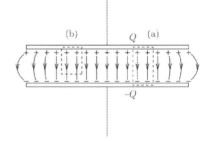

$$V = Ed = \frac{Qd}{\epsilon_0 S}. \tag{2.17}$$

Thus, the capacitance is determined to be

$$C = \frac{Q}{V} = \frac{\epsilon_0 S}{d}. \tag{2.18}$$

The unit of capacitance is [C/V] and is newly defined as [F] (**farad**). Hence, it is common to enlarge the area S and reduce the distance d, to obtain a large capacitance. It is also useful to fill the space with a dielectric material with a dielectric constant ϵ larger than ϵ_0.

We regard the concentric spherical conductors in Fig. 2.6 as a kind of capacitor and determine its capacitance. When we apply electric charge Q to the inner conductor and ground the outer conductor, the electric potential is given by Eq. (2.10). Hence, the electric potential difference between the two conductors is $V = (b - a)Q/(4\pi\epsilon_0 ab)$, and the capacitance is given by

$$C = \frac{4\pi\epsilon_0 ab}{b - a}. \tag{2.19}$$

When the distance between the electrodes, $d = b - a$, is sufficiently small, $4\pi ab$ is roughly equal to the area of the electrode S. Hence, it can be understood that this result agrees with the result for the parallel-plate capacitor, Eq. (2.18), in such a limit.

Example 2.4 Regard the concentric cylindrical conductors in Fig. 2.15 as a kind of capacitor. Determine its capacitance.

Fig. 2.15 Concentric cylindrical capacitor

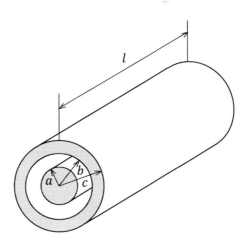

Solution 2.4 We add electric charges Q and $-Q$ to the inner and outer electrodes, respectively. These electric charges are uniformly distributed on the surfaces at $R = a$ and $R = b$, respectively. The electric field has a nonzero value only in the region $a < R < b$ and its strength is

$$E = \frac{Q}{2\pi \epsilon_0 R l}.$$

Hence, the electric potential difference between electrodes is

$$V = \frac{Q}{2\pi \epsilon_0 l} \int_a^b \frac{\mathrm{d}R}{R} = \frac{Q}{2\pi \epsilon_0 l} \log \frac{b}{a},$$

and the capacitance is determined to be

$$C = \frac{Q}{V} = \frac{2\pi \epsilon_0 l}{\log(b/a)}.$$

When the distance between the electrodes is sufficiently small that $\delta = 1 - b/a$ is much smaller than 1, the capacitance is approximately given by $C \simeq 2\pi a l \epsilon_0 / a\delta$, where we used the approximation, $\log(1 + \delta) \simeq \delta$. Since $2\pi a l$ and $a\delta$ represent the area and distance of the electrodes, respectively, this result corresponds to that in Eq. (2.18).

$\diamond$

2.4 Dielectric Materials

As described in Sect. 2.1, electrons are bonded to nuclei in a dielectric material, and hence, they cannot move freely to shield perfectly the material from applied electric field. The dielectric material is shielded to some extent, however, due to displacement of bonded electrons.

When an external electric field is applied to a dielectric material, bonded electrons are displaced in the opposite direction, and nuclei with positive charges are displaced in the direction of the applied electric field. Such a couple of positive and negative charges of equal magnitude that are displaced in the opposite directions to each other is called **electric dipole**. The quantity that expresses its strength quantitatively is **electric dipole moment**. When electric charges $\pm q$ are displaced and the position vector directed from the negative charge to the positive charge is d (see Fig. 2.16), the electric dipole moment is given by

Fig. 2.16 Electric dipole

$$p = qd. \qquad (2.20)$$

When many electric dipoles exist, the resultant electric dipole moment in a unit volume is called **electric polarization** and is denoted by P. Its unit is $[C/m^2]$. Such a phenomenon of movement of different electric charges with opposite directions is also called electric polarization.

When electric charges inside a dielectric material are displaced by an external electric field, as shown in Fig. 2.17a, the electric field inside the material is neutral because of cancellation, but some charges appear on the surfaces as shown in Fig. 2.17b. This is similar to the case of a conductor (see Fig. 2.2). The difference is that the electric charge on the surface is not true charge that can be transferred outside. This electric charge is called **polarization charge**. We denote the positive polarization charge density and the relative displacement by ρ_p and d, respectively. Then, the density of the positive polarization charge on the surface is $\sigma_p = \rho_p d$. On the other hand, the electric dipole in a unit volume is given by $P = \rho_p d$ from the definition, and hence,

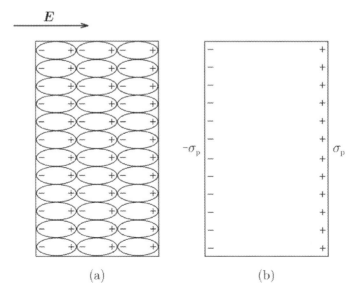

Fig. 2.17 a Mutual displacement of electric charges inside a dielectric material due to electric polarization and **b** resultant polarization charges on the surfaces

we have

$$P = \sigma_p. \tag{2.21}$$

Namely, the magnitude of electric polarization is equal to the polarization charge density on the surface for a uniform electric polarization.

When the electric polarization occurs, it produces an electric field. This is the same as that produced by true electric charge. If it is denoted by E_p, we have

$$E_p = -\frac{P}{\epsilon_0}. \tag{2.22}$$

Note that the electric polarization weakens the applied electric field. We denote by D/ϵ_0 the electric field strength produced only by true electric charge. Then, the total electric field is given by

$$E = \frac{D - P}{\epsilon_0}. \tag{2.23}$$

Then, D is given by

$$D = \epsilon_0 E + P \tag{2.24}$$

and is called the **electric flux density**. The unit of electric flux density is the same as that of electric polarization, $[C/m^2]$. In general, P is, and hence, D is also proportional to the electric field strength E. So, we can write it as

$$D = \epsilon E, \tag{2.25}$$

where ϵ is called the **dielectric constant**. When we write ϵ as

$$\epsilon = \epsilon_0 \epsilon_r, \tag{2.26}$$

ϵ_r is called the **relative dielectric constant**. The relative dielectric constants of various materials are listed in Table 2.1. Since D/ϵ_0 is the electric field strength produced by true electric charge, **Gauss's law**, which was given by Eq. (1.25), is written as

$$\int_S D \cdot dS = \int_V \rho(r) dV. \tag{2.27}$$

The surface integral of the electric flux density given by the left-hand side is called the **electric flux**. In other words, the electric charge density on the right-hand side of Eq. (1.25) includes the polarization charge.

Table 2.1 Values of relative dielectric constant for various materials at room temperature

Gas (1 atm)		Solid	
Oxygen	1.00049	Titanium dioxide	83–183
Nitrogen	1.00055	Quartz glass	3.5–4.5
Carbon dioxide	1.00091	Mica	5–9
Liquid		Ebonite	2.6–5.0
Water	78.54	Bakelite	4.5–9.0
Ethyl alcohol	24.30	Polyethylene	2.3–2.7
Solid		Vinyl chloride	3.3–6.0
Sodium chloride (NaCl)	5.9	Ferroelectric material	
Silicon (Si)	10.7–11.8	Barium titanate	1150–4500
Aluminum oxide (Al_2O_3)	8.5–11	Rochelle salt	∼ 4000

Example 2.5 Assume that electric field of strength E_0 is applied normal to the wide surface of a dielectric material with dielectric constant ϵ, as shown in Fig. 2.18. Determine the electric field strength and electric polarization in the dielectric material, and the polarization charge density on the surface.

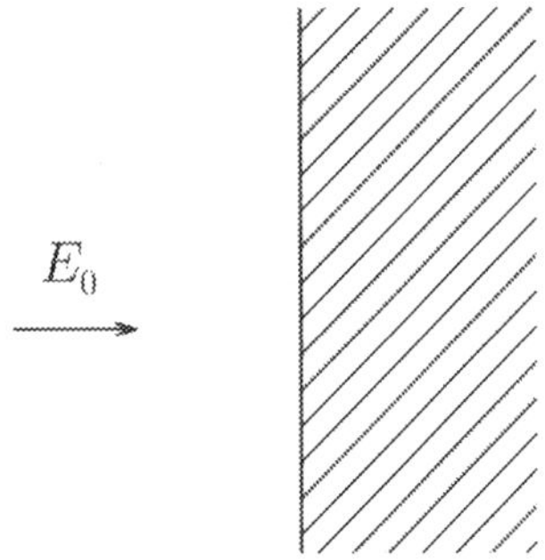

Fig. 2.18 Electric field strength E_0 applied normal to the wide surface of a dielectric material

E_0

Solution 2.5 Assume a closed surface S, as shown in Fig. 2.19a. We apply Gauss's law, Eq. (2.27), on S. Since true charge does not appear on the surface of the dielectric material, the normal component of the electric flux density is continuous between vacuum and the dielectric material. The electric flux density in vacuum is $\epsilon_0 E_0$. So, if we denote the electric field strength in the dielectric material by E, the above continuity leads to $\epsilon_0 E_0 = \epsilon E$. Thus, we have

$$E = \frac{\epsilon_0}{\epsilon} E_0. \tag{2.28}$$

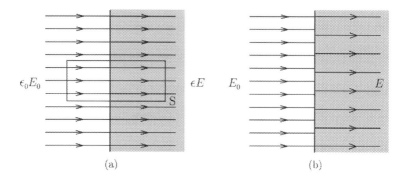

$\epsilon_0 E_0$ ϵE E_0 E

(a)   (b)

Fig. 2.19 Conditions of **a** the electric flux density and **b** the electric field at the surface

Since $\epsilon > \epsilon_0$, the electric field strength is weaker inside the dielectric material, and the inside of the material is weakly shielded (see Fig. 2.19b). Thus, the electric polarization is determined to be

$$P = (\epsilon - \epsilon_0)E = \frac{\epsilon_0(\epsilon - \epsilon_0)}{\epsilon} E_0.$$

From Eq. (2.21), the polarization charge density on the surface is $\sigma_p = -P$, taking account of the direction of the electric field. Note that ρ includes the polarization charge density in Eq. (1.25).

$\diamond$

From the result of Example 2.5, it is shown that the normal component of the electric flux density is continuous at the interface between materials with different dielectric constants, when there is no true charge on the interface. We denote by D_{1n} and D_{2n} the normal components of the electric flux density near the interface. Then, we have

$$D_{1n} = D_{2n}. \tag{2.29}$$

When a true charge exists on the interface, the continuity does not hold. Assume that true electric charge of surface density σ exists on the interface. Then, we have

$$\boldsymbol{n} \cdot (\boldsymbol{D}_1 - \boldsymbol{D}_2) = \sigma, \tag{2.30}$$

where $\boldsymbol{D}_1$ and $\boldsymbol{D}_2$ are the electric flux densities in materials 1 and 2, respectively, and $\boldsymbol{n}$ is the unit vector normal to the interface and directed from material 2 to material 1.

Here, we assume a closed rectangle C that contains the interface between two materials with different dielectric constants with both sides parallel to the interface, as shown Fig. 2.20. The electric field strength in each dielectric material is denoted by $\boldsymbol{E}_1$ and $\boldsymbol{E}_2$, respectively. If we apply Eq. (1.17) on C in the limit $h \rightarrow 0$, we have

Fig. 2.20 Closed rectangle
C at the interface between
materials with different
dielectric constants and
electric fields in the vicinity
of C

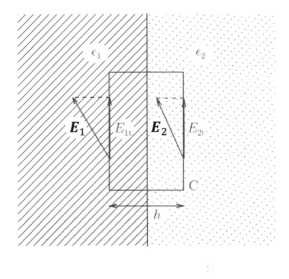

$$E_{1t} = E_{2t}, \tag{2.31}$$

where E_{1t} and E_{2t} are the parallel components of $\boldsymbol{E}_1$ and $\boldsymbol{E}_2$, respectively. Thus, the parallel component of the electric field is continuous on the interface. This condition is independent of whether there is true charge on the interface. Equation (2.31) can be generally expressed as

$$\boldsymbol{n} \times (\boldsymbol{E}_1 - \boldsymbol{E}_2) = 0. \tag{2.32}$$

Here, we discuss the refraction of electric field lines at the interface between different dielectric materials. Assume that electric field strength E_1 is applied in material 1 with dielectric constant ϵ_1 in the direction of angle θ_1 from the normal direction to the interface, as shown in Fig. 2.21. The electric field strength E_2 and its angle θ_2 in dielectric material 2 with ϵ_2 are determined using the boundary conditions. From the continuity of the parallel component of the electric field strength given by Eq. (2.31), we have

$$E_1\sin\theta_1 = E_2\sin\theta_2. \tag{2.33}$$

Equation (2.29) showing the continuity of the normal component of the electric flux density gives

$$\epsilon_1 E_1\cos\theta_1 = \epsilon_2 E_2\cos\theta_2. \tag{2.34}$$

From Eqs. (2.33) and (2.34), we have

Fig. 2.21 Refraction of
electric field lines at the
interface

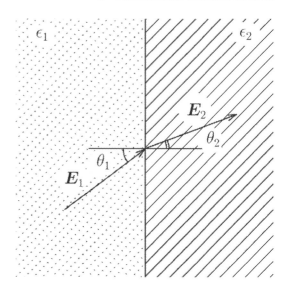

$$\frac{\tan\theta_1}{\tan\theta_2} = \frac{\epsilon_1}{\epsilon_2}. \tag{2.35}$$

This is the **law of refraction**. The value and angle of the electric field in material 2
are

$$E_2 = E_1 \left[\sin^2\theta_1 + \left(\frac{\epsilon_1}{\epsilon_2}\right)^2 \cos^2\theta_1 \right]^{1/2}, \tag{2.36}$$

$$\theta_2 = \tan^{-1}\left(\frac{\epsilon_2}{\epsilon_1}\tan\theta_1\right). \tag{2.37}$$

Thus, the refraction of the electric field can be explained using the boundary
conditions of electric field strength and electric flux density.

Suppose that the space between the electrodes in a parallel-plate capacitor in
Fig. 2.13 is fully occupied by a dielectric material with dielectric constant ϵ, and
that the electric field strength E_0 in Example 2.5 is produced by electric charges
$\pm Q$. Then, the electric potential difference between the electrodes is obtained from
Eq. (2.28) to be

$$V = \frac{\epsilon_0}{\epsilon} E_0 d = \frac{Qd}{\epsilon S}. \tag{2.38}$$

Hence, the capacitance of the capacitor is

$$C = \frac{\epsilon S}{d}. \tag{2.39}$$

Thus, from Eq. (2.18), the capacitance can be increased by using a dielectric material with a large dielectric constant.

Example 2.6 Two kinds of dielectric material with dielectric constants ϵ_1 and ϵ_2 occupy the space between the electrodes of a parallel-plate capacitor, as shown in Fig. 2.22. Determine the capacitance of the capacitor.

Fig. 2.22 Parallel-plate capacitor with two kinds of dielectric material

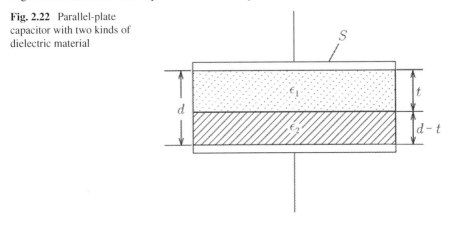

Solution 2.6 We give electric charges $\pm Q$ to each electrode. Then, the electric charge is uniformly distributed on the electrode surface and the charge density is $\sigma = Q/S$. Using Gauss's law, the electric flux density in dielectric materials 1 and 2 is $D_1 = D_2 = \sigma$. Hence, the electric field strength in each region is

$$E_1 = \frac{D_1}{\epsilon_1} = \frac{Q}{\epsilon_1 S}, \quad E_2 = \frac{D_2}{\epsilon_2} = \frac{Q}{\epsilon_2 S}.$$

The electric potential difference between the electrodes is determined to be

$$V = E_1 t + E_2 (d - t) = \frac{Q}{S}\left(\frac{t}{\epsilon_1} + \frac{d - t}{\epsilon_2}\right).$$

This yields the capacitance:

$$C = \frac{Q}{V} = \frac{\epsilon_1 \epsilon_2 S}{\epsilon_1 (d - t) + \epsilon_2 t}.$$

◇

Example 2.7 Two kinds of dielectric material with dielectric constants ϵ_1 and ϵ_2 occupy the space between the electrodes of a parallel-plate capacitor, as shown in Fig. 2.23. Determine the capacitance of the capacitor.

Fig. 2.23 Parallel-plate capacitor with two kinds of dielectric material

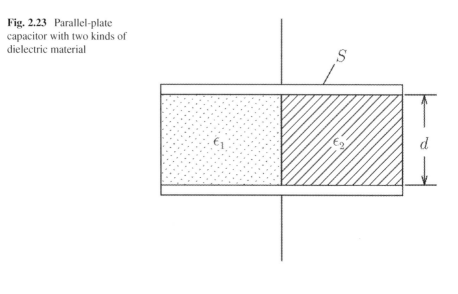

Solution 2.7 We apply voltage V to the capacitor. Then, the electric field strength in dielectric materials 1 and 2 is $E_1 = E_2 = V/d$. So, the electric flux density in each region is

$$D_1 = \frac{\epsilon_1 V}{d}, \quad D_2 = \frac{\epsilon_2 V}{d}.$$

Since the density of electric charge that appears on the surface of the electrode facing each dielectric material is

$$\sigma_1 = D_1, \quad \sigma_2 = D_2,$$

the total electric charge is determined to be

$$Q = (\sigma_1 + \sigma_1)\frac{S}{2} = \frac{(\epsilon_1 + \epsilon_2)SV}{2d}.$$

Thus, we have the capacitance:

$$C = \frac{Q}{V} = \frac{(\epsilon_1 + \epsilon_2)S}{2d}.$$

◇

When we apply a voltage to the capacitor, the electric field strength in each region cannot be obtained in Exercise 2.6. When we give an electric charge to the capacitor, the charge density on the surface of the electrode facing each material cannot be obtained in Exercise 2.7. Thus, we have used the above solution methods.

2.5 Electrostatic Energy

We suppose a process to transfer electric charge to a capacitor of capacitance C using a power source. When the output voltage of the power source is V', the electric charge of the capacitor is $Q' = CV'$. On increasing the output voltage of the power source by $\delta V'$ from this condition, the electric charge, $\delta Q' = C\delta V'$, is transferred to the capacitor. The work necessary to transfer this amount of electric charge is given from Eq. (1.16) by

$$\delta W = CV'\delta V'. \tag{2.40}$$

So, the work necessary while increasing the voltage from 0 to V is given by

$$W = \int_0^V CV'\mathrm{d}V' = \frac{1}{2}CV^2. \tag{2.41}$$

Since this is the work done from the condition of no electric charge, i.e., the condition of no electric energy, we can regard this as the **electrostatic energy** that charged capacitor has. Using the relationship of Eq. (2.18), the electrostatic energy can also be expressed as

$$U_e = \frac{1}{2}CV^2 = \frac{1}{2}QV = \frac{1}{2C}Q^2. \tag{2.42}$$

When the electrostatic energy of the parallel-plate capacitor in Fig. 2.13 is expressed using the corresponding electric field strength in Eq. (2.17), we have

$$U_e = \frac{1}{2}\epsilon_0 E^2 Sd, \tag{2.43}$$

where Sd is the volume of the space in which the electric field is concentrated and has a constant strength. So, we can consider

$$u_e = \frac{1}{2}\epsilon_0 E^2 \tag{2.44}$$

as the electrostatic energy in a unit volume. So, this called the **electrostatic energy density**. When a dielectric material of dielectric constant ϵ occupies the space between the electrodes, we can replace ϵ_0 in Eq. (2.44) by ϵ.

Example 2.8 Determine the electrostatic energy when electric charges $\pm Q$ are applied to the parallel-plate capacitor in Fig. 2.22. Derive the capacitance of this capacitor using this result.

Solution 2.8 Using Gauss's law, the electric field strength in each dielectric material is

$$E_1 = \frac{Q}{\epsilon_1 S}, \quad E_2 = \frac{Q}{\epsilon_2 S}.$$

So, the electrostatic energy density in each dielectric material is given by

$$u_{e1} = \frac{Q^2}{2\epsilon_1 S^2}, \quad u_{e2} = \frac{Q^2}{2\epsilon_2 S^2}.$$

The total electrostatic energy is determined to be

$$U_e = [u_{e1}t + u_{e2}(d - t)]S = \frac{Q^2}{2S}\left(\frac{t}{\epsilon_1} + \frac{d - t}{\epsilon_2}\right).$$

Using Eq. (2.42), we have the capacitance:

$$C = \frac{Q^2}{2U_e} = S\left(\frac{t}{\epsilon_1} + \frac{d - t}{\epsilon_2}\right)^{-1} = \frac{\epsilon_1\epsilon_2 S}{\epsilon_1(d - t) + \epsilon_2 t}.$$

This agrees with the result of Exercise 2.6.

$\diamondsuit$

Exercises

2.1 Suppose that $b = 2a$ and $c = 3a$ for concentric spherical conductors in Fig. 2.6. When we give a certain charge and charge Q to the inner and outer conductors, respectively, the electric potential at $r = 3a/2$ is zero. Determine the electric charge given to the inner conductor.

2.2 Prove that Eq. (2.4) is fulfilled on the conductor surface in Example 2.1.

Fig. 2.24 Electric field
strength E_0 applied parallel
to the wide surface of a
dielectric material

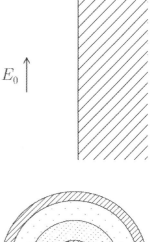

E_0 ↑

Fig. 2.25 Concentric
spherical capacitor with two
kinds of dielectric material

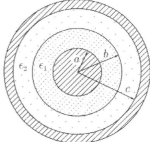

2.3 Suppose that electric charges Q and $2Q$ are given to the left and right slab
conductors, respectively, in Fig. 2.10. Determine the electric charge on each
surface.

2.4 The electric potential and electric field strength in vacuum in Fig. 2.11 are
obtained by the given electric charge and the image charge placed at the
symmetric point with respect to the conductor surface after virtually removing
the conductor. Determine the electric potential and electric field strength, and
prove Eq. (2.11) on the conductor surface.

2.5 Suppose that electric field of strength E_0 is applied parallel to the wide surface
of a dielectric material with dielectric constant ϵ, as shown in Fig. 2.24. Determine the electric field strength, electric flux density, and electric polarization
in the dielectric material, and the polarization charge density on the surface.

2.6 Determine the capacitance of the concentric spherical capacitor with two kinds
of dielectric material shown in Fig. 2.25.

2.7 Determine the capacitance of the concentric spherical capacitor with two kinds
of dielectric material shown in Fig. 2.26.

2.8 Suppose that we give electric charge Q to a spherical conductor of radius a.
Determine the electrostatic energy from the volume integral of Eq. (2.44).

2.9 Solve the problem in (8) using the following method. Regarding that the spherical conductor and infinity as electrodes of a capacitor, the electrostatic energy

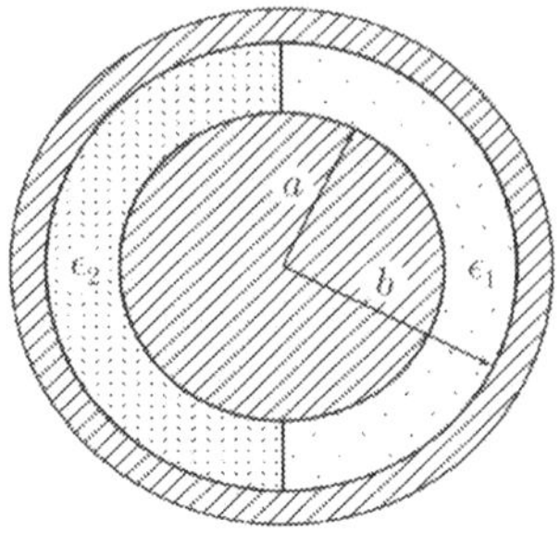

Fig. 2.26 Concentric
spherical capacitor with two
kinds of dielectric material

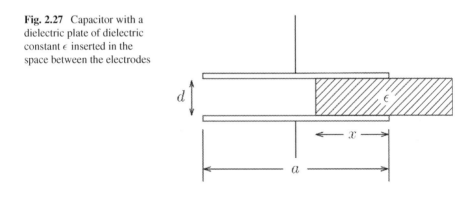

is given by $U_e = (1/2)Q\phi$ with the electric potential ϕ of the spherical
conductor.

2.10   Calculate the electrostatic energy of the capacitor when a dielectric plate with
dielectric constant ϵ is inserted in the space between the electrodes up to the
depth x from the edge, as shown in Fig. 2.27. The size of the electrodes in the
direction normal to the page is b. It is assumed that the capacitor has electric
charges $\pm Q$. Determine the force on the dielectric plate from the derivative of
the electrostatic energy with respect to x: $F = -\partial U_e/\partial x$.

Fig. 2.27 Capacitor with a
dielectric plate of dielectric
constant ϵ inserted in the
space between the electrodes

Chapter 3
Steady Current

Abstract This chapter covers electric phenomena when there is a steady current flow. When a current is applied to a substance, an electric potential difference appears, and there is a proportional relationship between the current and the electric potential difference. This relationship is known as Ohm's law on electric resistance. In this chapter, many examples of how to calculate resistance are shown. We learn also about the Joule loss that occurs when a current is applied to a substance with electric resistance. To realize a steady current in a closed circuit, we need an electric power source with an electromotive force. The fundamental principles on the behavior of electric circuits known as Kirchhoff's law are given by the relationship between the electromotive force and the potential drop due to electric resistance, and by the fundamental feature of steady current.

3.1 Current

There are innumerable electrons that can move freely in a conductor such as a metal. When we impose an electric potential difference on a conductor, the electric field drives electrons. Thus, the motion of electric charge, i.e., the **current**, occurs. To distinguish such a current from magnetizing current and displacement current, a current due to the motion of true charges is called **true current**.

Current is a vector with magnitude and direction. When electric charge dQ passes through a cross-section of the conductor within a small time-interval dt, its magnitude is given by

$$I = \frac{dQ}{dt}. \tag{3.1}$$

The unit of current is [C/s] and is newly defined as [A] (**ampere**). The quantity that expresses the strength of the current is the **current density** i. Its direction is the same as that of the current, and when current dI passes through a cross-sectional area dS perpendicular to the current direction, the magnitude is given by

Fig. 3.1 Electric charge
inside V and current going
out through dS on S

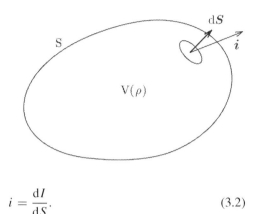

$$i = \frac{dI}{dS}. \tag{3.2}$$

The unit of current density is [A/m^2]. When the current density i is given, the current that passes through the small cross-section dS is

$$dI = i \cdot dS. \tag{3.3}$$

Suppose that electric charge is distributed with density ρ inside region V. The **principle of conservation of electric charge** means that the increase in the total electric charge in the region in unit time is equal to the inflow of the current into the region. This is expressed as

$$\frac{d}{dt} \int_V \rho(r)dV = - \int_S i \cdot dS, \tag{3.4}$$

where S is the surface of V, and dS is an elementary surface vector on S directed outward (see Fig. 3.1). In the steady state, there is no change in time. Hence, the left side of Eq. (3.4) is zero and we have

$$\int_S i \cdot dS = 0. \tag{3.5}$$

Such a current is called a **steady current**.

Example 3.1 Suppose that a capacitor of capacitance 3 μF is connected to a power source, and we apply a voltage of 5 V for 10 s. Determine the current from the power source in this period.

Solution 3.1 The electric charge stored during the period of applied voltage is 1.5×10^{-5} C from the relation $Q = CV$. Since the electric charge is transferred for 10 s, the current is $I = dQ/dt = 1.5\,\mu A$.

◇

3.2 Ohm's Law

It is necessary to apply a voltage to a material such as a metal so that a steady current flows in it. In this case, it is empirically known that a proportional relationship holds between the voltage V and the current I:

$$V = RI, \tag{3.6}$$

where the proportional constant R is called the **electric resistance** or simply **resistance**. Electric resistance is determined by the shape and electric properties of the material. This is called **Ohm's law**. The unit of electric resistance is [V/A] and is newly defined as [Ω] (**ohm**). When the length and the area of the uniform cross-sectional area are l and S, respectively, the electric resistance of the material is given by

$$R = \rho_r \frac{l}{S}, \tag{3.7}$$

where ρ_r is a constant inherent to material called the **resistivity** or **specific resistance**. Its unit is [Ωm]. Values of the resistivity of various materials are listed in Table 3.1. There is no other material constant that varies as widely as the resistivity, and the electric property changes dramatically due to the difference in the resistivity. Materials with the resistivity less than $10^{-6}\Omega$m that can easily transport a current are classified as conductors, those with resistivity above $10^{8}\Omega$m that can hardly do so are insulators, and those with intermediate resistivity are semiconductors. The constant,

$$\sigma_c = \frac{1}{\rho_r}, \tag{3.8}$$

Table 3.1 Resistivity of various materials at 20 °C

Metal	$(\times 10^{-8}\Omega\text{m})$	Semiconductor	(Ωm)
Silver (Ag)	1.62	Germanium (Ge)*	4.8×10^{-1}
Copper (Cu)	1.72	Silicon (Si)*	3.2×10^{3}
Gold (Au)	2.4	*Insulator*	(Ωm)
Aluminum (Al)	2.75	Epoxy resin	$10^{11} - 10^{14}$
Brus (Cu–Zn)	5–7	Aluminum oxide	$10^{12} - 10^{13}$
Iron (Fe)	9.8	Mica	$10^{12} - 10^{15}$
Platinum (Pt)	10.6	Natural rubber	$10^{13} - 10^{15}$
Constantan	50	Polyethylene	$> 10^{14}$
Mercury (Hg)	95.8	Paraffin	$10^{14} - 10^{17}$
Nichrome	109	Quartz glass	$> 10^{15}$

(*Values at 27 °C)

Fig. 3.2 Small region in
which current flows under an
electric potential difference

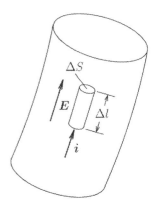

Fig. 3.2 Small region in
which current flows under an
electric potential difference

is called the **electric conductivity**. Its unit is [S/m], where S is **siemens**.

Suppose that there is a small region in which a current flows due to an electric field, as sown in Fig. 3.2. The length of this region along the current is Δl, and the cross-sectional area normal to the current is ΔS. When the electric field strength and the current density are E and i, respectively, the electric potential difference and the current in this region are $\Delta V = E \Delta l$ and $\Delta I = i \Delta S$, respectively. Hence, the electric resistance is

$$R = \frac{\Delta V}{\Delta I} = \frac{E}{i} \cdot \frac{\Delta l}{\Delta S}. \tag{3.9}$$

On the other hand, the electric resistance is defined as

$$R = \rho_{\mathrm{r}} \frac{\Delta l}{\Delta S}. \tag{3.10}$$

Thus, we have the local relationship: $E = \rho_{\mathrm{r}} i$. Since the directions are the same between the electric field and the current, this relationship is written as

$$\boldsymbol{E} = \rho_{\mathrm{r}} \boldsymbol{i} \tag{3.11}$$

or

$$\boldsymbol{i} = \sigma_{\mathrm{c}} \boldsymbol{E}. \tag{3.12}$$

These are also called **Ohm's law**.

Electric behavior in such materials is quite different from that in vacuum. Electric charges are accelerated by an electric field, and no steady state can occur in vacuum. In a material, on the other hand, electric charges move steadily without being accelerated in an electric field. That is, electric charges, mostly electrons, are deaccelerated every time that they collide with atoms in the material. As a result, if averaged for a longer time, the speed is approximately constant, and a steady state is realized. Even in such a case, the electric potential can be defined as in vacuum.

Example 3.2 Determine the electric resistance along the length of a slab of length l, thickness t, and width that varies, as shown in Fig. 3.3. The resistivity of the material is ρ_r.

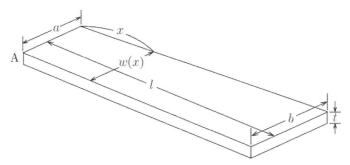

Fig. 3.3 Slab with width that varies along the length

Solution 3.2 The width at a distance x from edge A is

$$w(x) = a + \frac{b-a}{l}x.$$

When we apply current I to the slab, the current density at the distance x is given by

$$i(x) = \frac{I}{tw(x)}$$

and the electric field strength is $E(x) = \rho_r i(x)$. Thus, the voltage between the two edges is

$$V = \int_0^l E(x)\mathrm{d}x = \frac{\rho_r I}{t} \int_0^l \frac{\mathrm{d}x}{a + (b-a)x/l} = \frac{\rho_r l I}{t(b-a)} \log\frac{b}{a}.$$

The electric resistance is determined to be

$$R = \frac{V}{I} = \frac{\rho_r l}{t(b-a)} \log\frac{b}{a}.$$

$\diamond$

Example 3.3 Determine the electric resistance of a quarter of circle of radius r_0 with a rectangular cross-section, as shown in Fig. 3.4a. The resistivity is ρ_r.

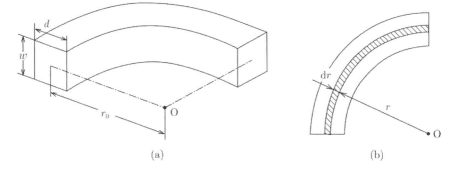

Fig. 3.4 **a** A quarter-circular prism and **b** part of a thin region with its radius from r to $r + dr$

Solution 3.3 We apply voltage V between the two edges. The electric field strength at an arc of radius r, as shown in Fig. 3.4b, is

$$E(r) = \frac{2V}{\pi r},$$

and the current density at this position is

$$i(r) = \frac{2V}{\pi \rho_r r}.$$

The current flowing in the region from r to $r + dr$ is $i(r)w\,dr$, and the total current is

$$I = \int_{r_0 - d/2}^{r_0 + d/2} \frac{2wV}{\pi \rho_r r}\,dr = \frac{2wV}{\pi \rho_r}\log\frac{r_0 + d/2}{r_0 - d/2}.$$

The electric resistance is determined to be

$$R = \frac{\pi \rho_r}{2w\log[(r_0 + d/2)/(r_0 - d/2)]}.$$

$\diamond$

3.3 Electromotive Force

Suppose that a steady current, I, flows in a closed circuit, C. When integrating the current along C, we have

$$\oint_C I \cdot ds = Il, \tag{3.13}$$

where I is the magnitude of I and l is the perimeter of the circuit. On the other hand, using Ohm's law, the left side of this equation leads to

$$S \oint_C i \cdot ds = S\sigma_c \oint_C E \cdot ds = 0, \tag{3.14}$$

where S and σ_c are the cross-sectional area and electric conductivity of the circuit, and Eq. (1.17) was used. Thus, we have $I = 0$, which contradicts the assumption of the steady current.

As can be seen easily, it is necessary to have an **electric power source** to realize a steady current. The electric potential difference generated by the electric power source is the **electromotive force**, and its unit is [V]. Practical electric power sources and kinds of electromotive force are listed in Table 3.2. The electric energy of the electromotive force provided by the source is transformed from chemical, mechanical, thermal, or optical energy. The electromotive force of electromagnetic induction, which is purely an electromagnetic phenomenon, as will be treated in Chap. 6, will be discussed from the viewpoint of electric circuits in Chap. 11. Here, we discuss a common electric power source that is not based on the electromagnetic induction and will treat only its output voltage.

Suppose that there is a closed circuit with an electric power source of electromotive force V_{em}, as shown in Fig. 3.5. We denote the part containing the electric power

Table 3.2 Kinds of electric power source and electromotive force

Electric power source	Kind of electromotive force
Battery	Chemical electromotive force
Generator	Electromagnetic induction
Thermocouple	Thermoelectric power
Photoelectric cell	Photovoltaic effect

Fig. 3.5 Closed electric circuit with electric power source

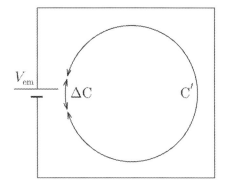

source and the remaining part of the electric circuit as ΔC and $C'(= C - \Delta C)$, respectively. The electric field due to the electromotive force is denoted by $\boldsymbol{E}_{\text{em}}$. Then, we have

$$V_{\text{em}} = - \int_{\Delta C} \boldsymbol{E}_{\text{em}} \cdot d\boldsymbol{s}, \tag{3.15}$$

where $d\boldsymbol{s}$ is directed along the current (see Fig. 3.6). We define the electric field, including that due to the electromotive force:

$$\boldsymbol{E} = \rho_r \boldsymbol{i} \tag{3.16}$$

in C', which does not include the electric power source and

$$\boldsymbol{E} = \boldsymbol{E}_{\text{em}} \tag{3.17}$$

in ΔC. From the condition that the electric field satisfies

$$\oint_C \boldsymbol{E} \cdot d\boldsymbol{s} = 0, \tag{3.18}$$

we have

$$\frac{\rho_r}{S} \int_{C'} \boldsymbol{I} \cdot d\boldsymbol{s} = V_{\text{em}}. \tag{3.19}$$

Thus, a steady current is realized by the electric power source. The electric potential difference on the left side is due to the current that flows through the resistance in the circuit and is called the **voltage drop**. The features discussed above give the

Fig. 3.6 Electric potential and direction of electric field in closed circuit

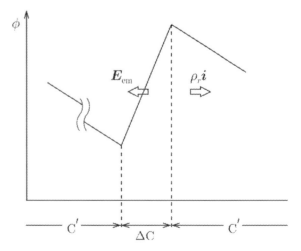

basis of Kirchhoff's law, which is s fundamental law of electric circuits and will be introduced in Chap. 11.

3.4 Joule Loss

When a current flows in a usual material with electric resistance by applying a voltage, electric charges are not accelerated, and there is a steady current flow. That is, the kinetic energy of the electric charges does not increase even when energy is supplied. So, the given energy is dissipated. Suppose that current I flows under a potential difference V between the two edges of a material of some length. The amount of electric charge that is transferred during a small period Δt is $\Delta Q = I \Delta t$. The work done in this period is given by Eq. (1.16):

$$\Delta W = V \Delta Q = V I \Delta t. \tag{3.20}$$

Hence, the work done, i.e., the energy dissipated, in unit time is

$$P = \frac{\Delta W}{\Delta t} = V I. \tag{3.21}$$

This is called the **electric power**. Its unit is [VA] and is defined as [W] (**watt**). The dissipated electric power changes to heat, and such a loss is called the **Joule loss**.

Using Ohm's law, the electric power is rewritten as

$$P = R I^2 = \frac{V^2}{R}. \tag{3.22}$$

We assume the case of a material of length l and uniform cross-section of area S. The dissipated electric power in a unit volume, i.e., the dissipated electric power density is given by

$$p = \frac{P}{lS} = \frac{V}{l} \cdot \frac{I}{S} = Ei. \tag{3.23}$$

Using the local Ohm's law, this is rewritten as

$$p = \sigma_c E^2 = \rho_r i^2. \tag{3.24}$$

Example 3.4 Determine the dissipated electric power density at the edge of width a when current I is applied to the slab in Example 3.3. By how much is it larger than the dissipated electric power density at the other edge of width b?

Solution 3.4 The current density at the edge of width a is

$$i(x = a) = \frac{I}{ta}.$$

From Eq. (3.24), the dissipated electric power density there is

$$p = \frac{\rho_r I^2}{(ta)^2}.$$

The dissipated electric power density at the other edge is similarly obtained as $p = \rho_r I^2/(tb)^2$. So, the former is $(b/a)^2$ times as large as the latter.

◇

Example 3.5 Determine the total dissipated electric power when current I is applied to the slab in Example 3.2.

Solution 3.5 The dissipated electric power density at a position at distance x from the edge A is

$$p(x) = \rho_r i(x)^2 = \rho_r \frac{I^2}{t^2 w(x)^2}.$$

Integrating this over the whole slab, we have

$$P = \int_0^l p(x)tw(x)dx = \frac{\rho_r I^2}{t} \int_0^l \frac{dx}{a + (b-a)x/l} = \frac{\rho_r l I^2}{t(b-a)} \log\frac{b}{a}.$$

This is rewritten as RI^2 using the resistance R obtained in Example 3.2.

◇

Exercises

3.1 We apply 3 μA for 20 s to a capacitor of capacitance 1 μF that has no electric charge. Determine the voltage between the two electrodes of this capacitor.

3.2 When the width b approaches a in Example 3.2, determine the resistance.

3.3 How does the resistance change when the width d is much smaller than r_0 in Example 3.3?

3.4 Determine the resistance along the length of the truncated corn in Fig. 3.7. The resistivity is ρ_r.

Fig. 3.7 Long truncated
corn

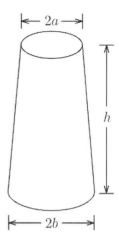

3.5   Suppose that the space $a \leq r \leq b$ of the concentric spherical conductors in
       Fig. 3.8 is occupied by a material with resistivity ρ_r. Determine the resistance
       between the two electrodes.

3.6   When we apply a current to the material in Example 3.3, by how much is the
       dissipated power density different between the innermost and outermost parts?

3.7   Prove that the dissipated electric power is equal to the commonly known result,
       $P = IV$, when electric voltage V is applied to the resistor shown in Fig. 3.4
       and current I flows.

3.8   Determine the dissipated electric power when voltage V is applied to the
       truncated corn shown in Fig. 3.7.

Fig. 3.8 Concentric
spherical resistors with the
space occupied by a material
with resistivity ρ_r

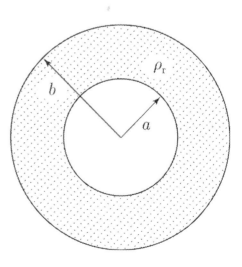

Chapter 4
Current and Magnetic Phenomena

Abstract This chapter covers magnetic phenomena in vacuum due to current. The Lorentz force works between currents. This magnetic reaction is considered to be caused by a magnetic distortion in space that is produced by other currents and this distortion is called magnetic field. The magnetic flux density is used as a quantity that expresses the strength of the magnetic field. The local magnetic flux density produced by currents is described by the Biot-Savart law. On the other hand, the global relationship between the current and the magnetic flux density is described by Ampere's law. This law is sometimes useful to determine the magnetic flux density.

4.1 Magnetic Phenomena Due to Current

The Lorentz force acts between currents. This is similar to the Coulomb force between electric charges. Hence, we can presume that currents also produce some field in space similar to the electric field produced by electric charges. This field is the **magnetic field**. Although the magnetic field can also be produced by magnets rather than currents, only currents can be quantitatively defined, and the magnetic interaction due to currents is described here. We use the **magnetic flux density** to express the strength of the magnetic field instead of the magnetic field itself. The magnetic field will be defined in Chap. 5.

When currents I_1 and I_2 flow on two parallel straight lines separated by distance d, as shown in Fig. 4.1a, a force of strength

$$F' = -\frac{\mu_0 I_1 I_2}{2\pi d} \tag{4.1}$$

works on each line of a unit length. In the above,

$$\mu_0 = 4\pi \times 10^{-7} \text{N/A}^2 \tag{4.2}$$

is the **magnetic permeability** of vacuum. The force is attractive ($F' < 0$) for currents in the same direction ($I_1 I_2 > 0$) and repulsive for currents in opposite directions. The magnitude of this force corresponds to that of the Coulomb force between linear

T. Matsushita, *Electricity*,
https://doi.org/10.1007/978-3-031-44002-1_4

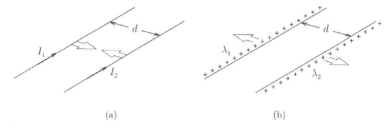

(a)   (b)

Fig. 4.1 **a** Force between two parallel currents and **b** force between two parallel line charges

parallel electric charges, as shown in Fig. 4.1b. That is, when electric charges of linear densities λ_1 and λ_2 are uniformly distributed on two parallel straight lines separated by d, the Coulomb force on each line of a unit length is given by

$$F' = \frac{\lambda_1 \lambda_2}{2\pi \epsilon_0 d}. \tag{4.3}$$

The unique difference is the difference of the directions of the forces.

4.2   The Biot-Savart Law

The conditions of magnetic flux lines can be visualized using magnetic particles such as iron filings. The structure of magnetic flux lines produced by a straight current is schematically shown in Fig. 4.2. The magnetic flux density vector is on a plane normal to the current, perpendicular to the radial vector, and forms vortices around it. This is completely different from the condition of the electric field shown in Fig. 1.5. This difference is related to the fact that the source of the field, i.e., the current, is a vector, while the source of the electric field, i.e., the electric charge, is a scalar.

Fig. 4.2 Magnetic flux density produced by a straight current

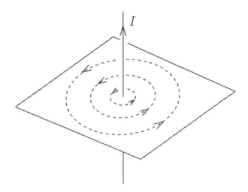

The magnetic flux density due to currents is described by the **Biot-Savart law**. Suppose that current flows along line C, as shown in Fig. 4.3. The magnetic flux density at point P at r produced by an elementary current $I d\mathbf{r}'$ flowing in a small segment $d\mathbf{r}'$ is given by

$$d\mathbf{B} = \frac{\mu_0 I d\mathbf{r}' \times (\mathbf{r} - \mathbf{r}')}{4\pi |\mathbf{r} - \mathbf{r}'|^3}. \tag{4.4}$$

Here, $\mathbf{r} - \mathbf{r}'$ is the position vector from the small segment to point P. The unit of the magnetic flux density is [T] (**tesla**). We denote $|\mathbf{r} - \mathbf{r}'| = r$ and $|d\mathbf{r}'| = dr'$, and by θ the angle from $d\mathbf{r}'$ to $\mathbf{r} - \mathbf{r}'$. Then, the magnitude of the magnetic flux density is

$$dB = \frac{\mu_0 I dr'}{4\pi r^2} \sin\theta \tag{4.5}$$

and the vector points along the motion of a screw when the screw driver is rotated from $d\mathbf{r}'$ to $\mathbf{r} - \mathbf{r}'$. That is, the direction of the magnetic flux density is perpendicular to both the current ($d\mathbf{r}'$) and the position vector ($\mathbf{r} - \mathbf{r}'$), as shown in Fig. 4.2. It is normal to the sheet and directed backward (into the sheet) in Fig. 4.3.

Hence, the magnetic flux density at point P at position r produced by current I flowing in line C is

$$\mathbf{B}(\mathbf{r}) = \frac{\mu_0}{4\pi} \int_C \frac{I d\mathbf{r}' \times (\mathbf{r} - \mathbf{r}')}{|\mathbf{r} - \mathbf{r}'|^3}. \tag{4.6}$$

When there are many currents, the total magnetic flux density is the superposition of all the individual magnetic flux densities they produce. When current flows with density i in space V, the magnetic flux density is given by

Fig. 4.3 Magnetic flux density at point P produced by elementary current $I d\mathbf{r}'$

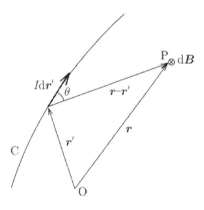

$$B(r) = \frac{\mu_0}{4\pi} \int_V \frac{i(r') \times (r - r')}{|r - r'|^3} dV', \tag{4.7}$$

where the integral is a volume integral with respect to r'.

Magnetic flux lines are defined for the magnetic flux density similarly to the electric field lines for the electric field. Since a magnetic flux line is a closed line, as can be seen from the example in Fig. 4.2, when we integrate on an arbitrary closed surface S, it is expected that we have

$$\int_S B \cdot dS = 0. \tag{4.8}$$

This is because any magnetic flux line that goes into a closed surface surely goes out from another position, since the line never terminates halfway. In fact, Eq. (4.8) is proven to be valid. This is a different point from the electric field lines given by Eq. (1.25).

Example 4.1 Determine the magnetic flux density at point A at distance a from a straight current I.

Solution 4.1 We define the x-axis along the direction of the current with the origin $x = 0$ at the foot of a perpendicular line from point A, as shown in Fig. 4.4. The magnetic flux density at point A produced by an elementary current $I\,dx$ in the region from x to $x + dx$ is

$$dB = \frac{\mu_0 I\,dx}{4\pi(x^2 + a^2)}\sin\theta.$$

Using the angle θ in Fig. 4.4, we have $x = -a\cot\theta$ and $dx = a\,d\theta/\sin^2\theta$. Thus, the above equation is written as

Fig. 4.4 Straight current I
and observation point A

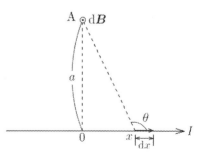

$$dB = \frac{\mu_0 I}{4\pi a} \sin\theta d\theta.$$

Since the magnetic flux density produced by each elementary current is directed forward normal to the sheet (out of the sheet), the total magnetic flux density is simply summed and we have

$$B = \frac{\mu_0 I}{4\pi a} \int_0^\pi \sin\theta d\theta = \frac{\mu_0 I}{2\pi a}. \tag{4.9}$$

◇

Example 4.2 Suppose that current I flows in a closed circuit composed of two quarter circles and two straight lines on a common plane, as shown in Fig. 4.5. Determine the magnetic flux density at point O.

Fig. 4.5 Current I flowing in a closed circuit

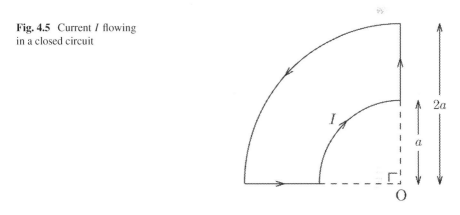

Solution 4.2 Since the angle θ in Eq. (4.5) is 0 and π for the straight sections, there is no contribution to the magnetic flux density from these sections. On the quarter circle of radius a, θ is $\pi/2$ and the magnetic flux density is directed backward normal to the sheet. Its contribution is

$$B_a = \frac{\mu_0 I}{4\pi a^2} \int dr' = \frac{\mu_0 I}{8a}.$$

On the quarter circle of radius $2a$, θ is $-\pi/2$ and the magnetic flux density is directed forward normal to the sheet. Its contribution is

$$B_{2a} = \frac{\mu_0 I}{4\pi (2a)^2} \int dr' = \frac{\mu_0 I}{16a}.$$

Thus, the magnetic flux density has a magnitude

$$B = B_a - B_{2a} = \frac{\mu_0 I}{16a}$$

and is directed backward normal to the sheet.

$\diamond$

4.3   Forces on Currents

The force on a current can be regarded as a force due to a magnetic distortion, i.e., the magnetic flux density in space, similarly to the Coulomb force on electric charges. The force on an elementary current $I\mathrm{d}s$ in the magnetic flux density $\boldsymbol{B}$ is

$$\mathrm{d}\boldsymbol{F} = I\mathrm{d}s \times \boldsymbol{B}. \tag{4.10}$$

This is called the **Lorentz force** and this relation is known as **Fleming's left-hand rule**. Hence, the force on current I flowing along C is

$$\boldsymbol{F} = I \int_C \mathrm{d}s \times \boldsymbol{B}. \tag{4.11}$$

Here, we shall determine the force between two parallel straight currents in Fig. 4.1a. The magnetic flux density that current I_1 produces at the position of current I_2 is $B = \mu_0 I_1/(2\pi d)$ and its direction is shown in Fig. 4.6. Hence, the force on current I_2 in a unit length is given by Eq. (4.1) and is attractive for $I_1 I_2 > 0$.

Current is a flow of particles with electric charges. If we denote the electric charge, number density, and velocity of a particle by q, n, and $\boldsymbol{v}$, respectively, the current density is written as

Fig. 4.6 Magnetic flux density and force on straight current I_2 produced by straight current I_1

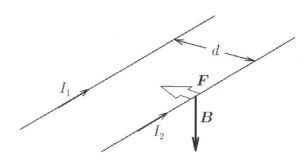

$$i = nq\boldsymbol{v}. \tag{4.12}$$

Hence, the force on a region comprising the cross-sectional area S and length l in which the current flows is

$$\boldsymbol{F} = Sl\boldsymbol{i} \times \boldsymbol{B} = nSlq\boldsymbol{v} \times \boldsymbol{B}. \tag{4.13}$$

In the above Sl is the volume of this region, and it multiplied by n, which is the number of particles in this volume. Hence, the Lorentz force on each particle is given by

$$\boldsymbol{f} = q\boldsymbol{v} \times \boldsymbol{B}. \tag{4.14}$$

Example 4.3 Two parallel currents flow in the same direction with distance $2a$, as shown in Fig. 4.7. Determine the magnetic flux density at point A, which is equidistant from each current. Determine also the force in a unit length and its direction on another current I' that flows at A.

Fig. 4.7 Two parallel currents flowing in the same direction and point A equidistant from each current

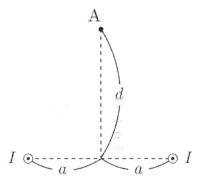

Solution 4.3 From Example 4.1, the magnetic flux density produced by one current is given by $B' = \mu_0 I / [2\pi (a^2 + d^2)^{1/2}]$. Since the vertical component is cancelled due to symmetry, as shown in Fig. 4.8, only the horizontal component remains. Hence, the magnetic flux density at point A is

$$B = 2B' \frac{d}{(a^2 + d^2)^{1/2}} = \frac{\mu_0 I d}{\pi (a^2 + d^2)}.$$

From Eq. (4.10), the force in a unit length on current I' flowing in the same direction at point A is given by

Fig. 4.8 Combined
magnetic flux density due to
two currents

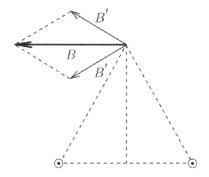

$$F' = I'B = \frac{\mu_0 I I' d}{\pi (a^2 + d^2)}$$

and the force is directed vertically downward.

◇

4.4   Ampere's Law

Suppose a circle C with its center on straight current I on a plane normal to the current, as shown in Fig. 4.9. We shall determine a circular integral of the magnetic flux density on it:

$$\oint_C \boldsymbol{B} \cdot \mathrm{d}\boldsymbol{s}. \tag{4.15}$$

Fig. 4.9 Straight current
and circle with the center on
the current

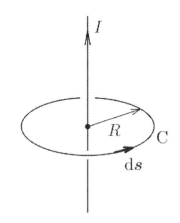

Since $\boldsymbol{B}$ is parallel to $d\boldsymbol{s}$, we have simply $\boldsymbol{B} \cdot d\boldsymbol{s} = Bds$. In addition, the value of B is given by $B = \mu_0 I/(2\pi R)$ from Eq. (4.9). Thus, we obtain

$$\oint_C \boldsymbol{B} \cdot d\boldsymbol{s} = \mu_0 I. \tag{4.16}$$

This result does not depend on the radius of the circle. In addition, the result is the same, even if C is not a circle but an arbitrary circuit that includes the current. When the current does not penetrate C but flows outside C, the circular integral is zero. This is called **Ampere's law**, and corresponds to Gauss's law on the electric field produced by electric charges.

When current flows with density $\boldsymbol{i}$, Ampere's law is written as

$$\oint_C \boldsymbol{B} \cdot d\boldsymbol{s} = \mu_0 \int_S \boldsymbol{i} \cdot d\boldsymbol{S}, \tag{4.17}$$

where S is the surface surrounded by C, and $d\boldsymbol{s}$ and $d\boldsymbol{S}$ follow the right-hand rule. That is, when a screw driver is rotated to the direction of $d\boldsymbol{s}$, $d\boldsymbol{S}$ is directed in the direction of motion of a screw.

Here, we show an example of determination of the magnetic flux density using Ampere's law. Suppose that a current of density i flows uniformly along a long cylindrical conductor of radius a. In this case, we can easily calculate the magnetic flux density using Ampere's law, while it is not easy to use the Biot-Savart law. First, we shall determine the magnetic flux density at a point of distance $r(< a)$ from the center of the cylinder. Suppose a virtual circle C of radius r on which the observation point is located, as shown in Fig. 4.10a. When Ampere's law is applied on C, the magnetic flux density has only the azimuthal component and takes on a constant value. Hence, the left side of Eq. (4.17) leads to

$$\int_C Bds = 2\pi r B. \tag{4.18}$$

On the other hand, the total current through C is $\pi r^2 i$ on the right side. Thus, the magnetic flux density is determined to be

$$B = \frac{\mu_0 r i}{2}. \tag{4.19}$$

Next, we shall determine the magnetic flux density outside the cylinder. We apply Ampere's law on a virtual circuit of radius $r(> a)$ with the origin on the central axis. The left side of Eq. (4.17) is the same as Eq. (4.18). The total current through C is $\pi a^2 i$. Thus, we have

$$B = \frac{\mu_0 a^2 i}{2r}. \tag{4.20}$$

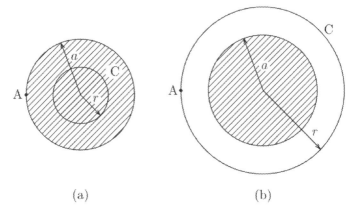

(a)   (b)

Fig. 4.10 Method to determine the magnetic flux density when current flows uniformly in a cylindrical conductor of radius a using Ampere's law for the magnetic field **a** inside the cylinder and **b** outside the cylinder

If we use the total current, $I = \pi a^2 i$, it is found that the result of Eq. (4.20) is the same as that of Eq. (4.9). That is, the magnetic flux density is the same as that in the case where all the current is concentrated on the center. The distribution of the magnetic flux density is shown in Fig. 4.11.

Fig. 4.11 Distribution of magnetic flux density when current flows in a cylindrical conductor

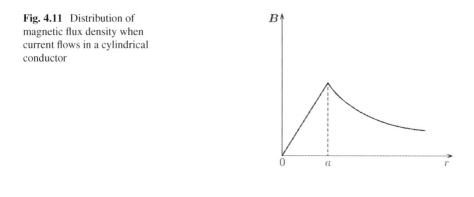

Example 4.4 Suppose that current flows uniformly with surface density τ [A/m] on a wide plane. Determine the magnetic flux density at point A at distance a from the plane.

Solution 4.4 Suppose a rectangle C that includes point A with two sides of length l parallel to the plane and the other sides normal to the plane, as shown in Fig. 4.12. We apply Ampere's law on C. We follow the right-hand rule for the direction of the

circular integral with respect to the current direction. The direction of the magnetic
flux density is parallel to the plane, and hence, to the two sides. Hence, it is perpendicular to the other two sides, and there is no contribution to the integral from these
sides. Since the distance from the plane is the same for the two parallel lines, the
magnitude of the magnetic flux density is the same, which is denoted by B. Then,
the left side of Eq. (4.17) is $2lB$.

On the other hand, the current penetrating C is τl. So, from Eq. (4.17) we have

$$B = \frac{\mu_0 \tau}{2},$$

which is independent of the distance a from the plane. This is similar to the electric
field strength produced by electric charge distributed uniformly on a wide surface
(see Example 1.5).

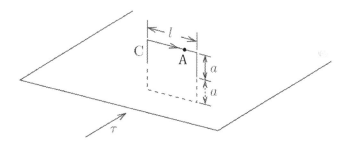

Fig. 4.12 Plane with uniform current and rectangle C

◇

Example 4.5 Solve the problem in Example 4.4 using the Biot-Savart law.

Solution 4.5 We define the x-axis in the direction normal to the current on the plane
with the origin $x = 0$ at the foot of a perpendicular line from the observation point.
The magnetic flux density at the observation point A produced by a linear current in
the region from x to $x + \mathrm{d}x$ is $\mathrm{d}B = \mu_0 \tau \mathrm{d}x / 2\pi \left(x^2 + a^2\right)^{1/2}$, and only the horizontal
component remains without cancellation from symmetry, as shown in Fig. 4.13:

$$\mathrm{d}B' = \mathrm{d}B\cos\theta = \frac{\mu_0 \tau a \mathrm{d}x}{2\pi \left(x^2 + a^2\right)}.$$

Using the relationship, $x = a\tan\theta$, the magnetic flux density is determined to be

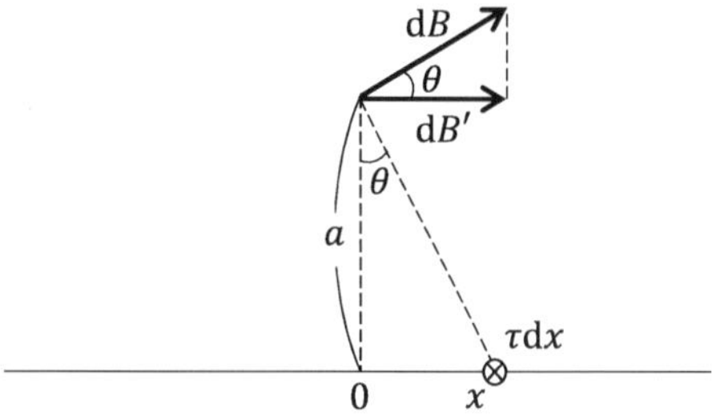

Fig. 4.13 Magnetic flux
density produced by a linear
current in the region from x
to $x + dx$

$$B = \int dB' = \frac{\mu_0 \tau}{2\pi} \int_{-\pi/2}^{\pi/2} d\theta = \frac{\mu_0 \tau}{2},$$

which agrees with the result of Example 4.4.

◇

Example 4.6 When current I flows uniformly on a long thin plate of width $2w$, as shown in Fig. 4.14, determine the magnetic flux density at point P at a distance $d(> w)$ from the center of the plane.

Fig. 4.14 Long thin plate
with current and observation
point P

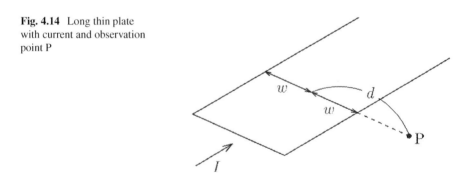

Solution 4.6 The magnetic flux density at P produced by a linear current $dI = (I/2w)dx$ in the region from x to $x + dx$ from the center is given by

$$dB = \frac{\mu_0 I dx}{4\pi w(d - x)}$$

and is directed normal to the plane, as shown in Fig. 4.15. Hence, if we divide the plane current into thin line currents, the contribution from each line current can be simply added. Thus, the magnetic flux density is determined to be

$$B = \int_{-w}^{w} \frac{\mu_0 I \, dx}{4\pi w(d-x)} = \frac{\mu_0 I}{4\pi w} \log \frac{d+w}{d-w}.$$

Fig. 4.15 Magnetic flux density produced by a linear current in the plane

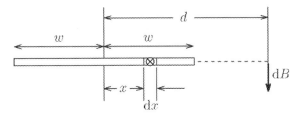

Exercises

4.1 Assume that currents flow in parallel four straight lines, as shown in Fig. 4.16. Determine the magnetic flux density at the center O.

4.2 Straight current I and unknown straight currents I_x and I_y flow parallel at positions placed at each vertex of the equilateral triangle with side length a, as shown in Fig. 4.17. The magnetic flux density at point A, which is symmetric with the position of the current I_x, is zero. Determine the values of I_x and I_y.

4.3 Two parallel currents flow in the opposite directions with distance $2a$, as shown in Fig. 4.18. Determine the force on current I' equidistant from each current.

4.4 Current I flows in a line composed of a semicircle of radius a and two straight lines on a common plane, as shown in Fig. 4.19. Determine the magnetic flux density at the center O of curvature of the semicircle.

4.5 Current I flows in a circle of radius a, as shown in Fig. 4.20. Determine the magnetic flux density at point P at a distance b in the normal direction from the center of the circle.

Fig. 4.16 Four parallel straight lines with current

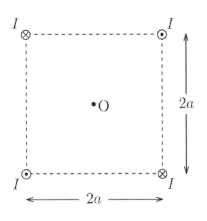

Fig. 4.17 Current I and unknown currents I_x and I_y

Fig. 4.18 Two parallel currents I flowing in the opposite directions and current I' equidistant from each current

Fig. 4.19 Current flowing in line composed of a semicircle and two straight lines

Fig. 4.20 Circular current

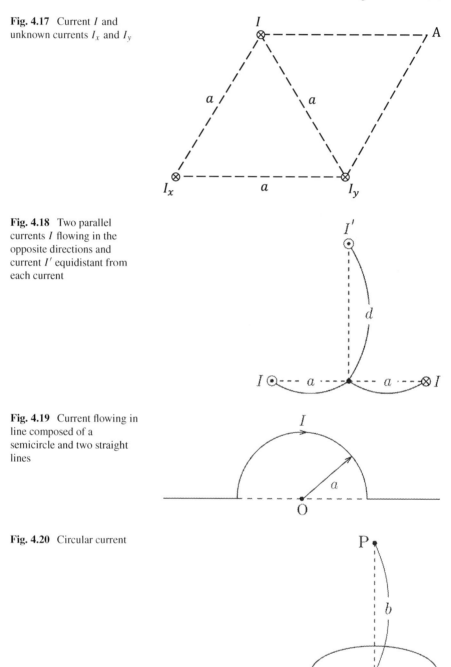

Fig. 4.21 Triangle in which current I flows and observation point A

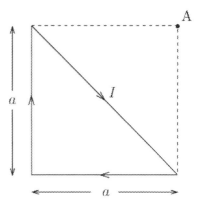

4.6 Current I flows in a closed triangle, as shown in Fig. 4.21. Determine the magnetic flux density at point A.

4.7 Current I flows in a closed square of side length a, as shown in Fig. 4.22. Determine the magnetic flux density at point P at a distance b from the center O of the square.

4.8 Current I flows uniformly on a long thin plate of width $2w$, as shown in Fig. 4.23. Determine the magnetic flux density at point P at a distance d from the center of the plane.

Fig. 4.22 Square current and observation point P

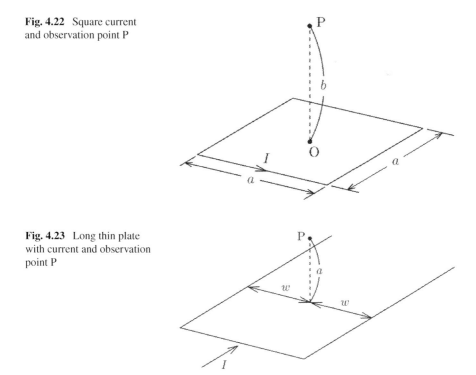

Fig. 4.23 Long thin plate with current and observation point P

Fig. 4.24 Current flowing
uniformly in a hollow
cylinder

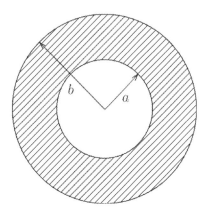

4.9   Current I flows uniformly in a hollow cylinder with the cross-section in
      Fig. 4.24. Determine the magnetic flux density in each region.

4.10 Current flows uniformly with density i normally forward to the page in a long
      cylinder of radius b with a hollow of radius a, as shown in Fig. 4.25. Determine
      the magnetic flux density at point P at a distance d from the central axis of
      the cylinder. (*Hint*: The same condition is obtained by superposing the case in
      which current flows with density i in the whole region including the hollow
      and the case in which current flows with density $-i$ in the hollow region.)

Fig. 4.25 Current flowing
uniformly with density i in a
cylinder with a hollow

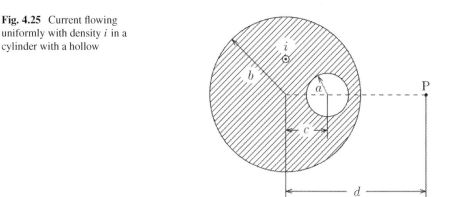

Chapter 5
Superconductors and Magnetic Materials

Abstract This chapter covers magnetic phenomena in magnetic materials, which are classified into superconductors and magnetic materials in the narrow sense. When a superconductor is placed in a magnetic flux density, the magnetic flux density is zero inside the superconductor due to a current flowing on its surface, similarly to electrical phenomena in a conductor in an electric field. On the other hand, the internal magnetic flux density is increased due to an effect of the internal magnetic moment for a magnetic material in a given magnetic flux density. The magnetic field is defined as a quantity that corresponds to a magnetic flux density produced only by true current. The refraction of magnetic flux lines at an interface between different magnetic materials is explained. Coils made of such magnetic materials are introduced, and we learn the inductance of coils, which is the magnetic flux that can be stored by unit current, 1 A.

5.1 Superconductors

Magnetic materials are classified into **superconductors** and **magnetic materials** in the narrow sense, based on their magnetic behavior, which is similar with the classification of electric materials into conductors and dielectric materials (insulators). For a superconductor in the **Meissner state**, i.e., a perfect diamagnetic state, the state of

$$\boldsymbol{B} = 0 \tag{5.1}$$

is maintained inside the superconductor, even when an external magnetic flux density is applied. So, from Eq. (4.17) it can be shown that

$$i = 0 \tag{5.2}$$

inside the superconductor. Such magnetic behavior is quite similar to the electric behavior of conductors represented by Eqs. (2.1) and (2.2). The perfect diamagnetic state given by Eq. (5.1) originates from true current flowing on the superconductor surface. This current is called the **Meissner current**.

T. Matsushita, *Electricity*,
https://doi.org/10.1007/978-3-031-44002-1_5

Here, we assume that current is applied to a superconductor. When current of surface density τ is applied along the y-axis on the surface $x = 0$ of the superconductor that occupies $x \geq 0$, as shown in Fig. 5.1, the magnetic flux density appears in a parallel direction to the superconductor surface. Here, we assume a rectangle C with sides parallel to the superconductor surface, and apply Eq. (4.17) on it. If we denote by l and B the length of the side of the rectangle along the y-axis and the magnetic flux density on the surface, respectively, the left side is given by $\mu_0 \tau l$. Hence, we have [1]

$$B = \mu_0 \tau. \tag{5.3}$$

When an external magnetic flux density B is applied parallel to a superconductor surface, the surface current of the density given by Eq. (5.3) flows and the interior of the superconductor is shielded. This relationship corresponds to the relationship of (2.4) for a conductor.

Thus, it is a common point between the conductor and superconductor that the source which shields the interior is true electric charge and its flow, i.e., true current, both of which can be transferred outside. The perfect diamagnetism of superconductor can be described with the vector potential (see Appendix A.2).

It is well known that the electric resistivity of superconductor is zero. Here, we shall prove it. Assume that a current is applied to the superconductor. Then, the current flows on the surface of the superconductor as shown above. We suppose a rectangle C located inside the superconductor, where one side is placed on the surface and parallel to the current, as shown in Fig. 5.2. When the current is integrated along C, the integral is not zero, while the current is zero on the three sides inside the superconductor. Thus, we have

$$\oint_C \boldsymbol{i} \cdot \mathrm{d}\boldsymbol{s} \neq 0. \tag{5.4}$$

Fig. 5.1 Surface current on the superconductor surface and magnetic flux density outside the superconductor

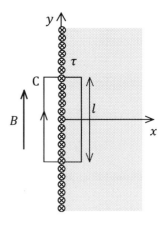

Fig. 5.2 Rectangle C in
superconductor with one side
on the surface on which
current flows

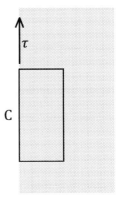

If Ohm's law holds, multiplication by the resistivity ρ_r leads to

$$\oint_C \boldsymbol{E} \cdot d\boldsymbol{s} \neq 0. \tag{5.5}$$

This contradicts Eq. (1.17), which describes the fundamental property of the electrostatic field. So, the zero resistivity of a superconductor can be proved [1].

Here, suppose that current I is applied to a long superconducting cylinder of radius a. We shall treat the magnetic flux density around the superconductor. Note that this is different from the case treated in Sect. 4.4, in which we apply the current to a usual long cylinder. The current does not flow uniformly but flows only on the surface of the superconductor. Assume circle C of radius r perpendicular to the central axis with the center on the axis, as shown in Fig. 5.3. We determine the magnetic flux density at point A on the circle using Ampere's law. The left side of Eq. (4.17) is equal to that of Eq. (4.18). Since the current that penetrates C is I, the right side of Eq. (4.17) is $\mu_0 I$. Thus, we have

$$B = \frac{\mu_0 I}{2\pi r}. \tag{5.6}$$

Fig. 5.3 Cross-sectional
view of the superconducting
cylinder in which current I
flows and circle C with the
center on the central axis

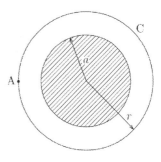

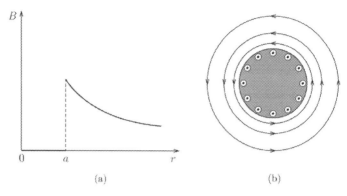

Fig. 5.4 a Magnetic flux distribution around the superconductor in which current *I* flows and **b** magnetic flux lines outside the superconductor

The magnetic flux density outside the superconductor is the same as in the case in which all the current is concentrated on the central axis. It is easily derived that the magnetic flux density inside the superconductor ($r < a$) is zero. The magnetic flux distribution around the superconductor is shown in Fig. 5.4.

The value of the magnetic flux density on the superconductor surface is $\mu_0 I / 2\pi a$ and the surface current density there is $\tau = I/2\pi a$. So, the relationship of Eq. (5.3) holds.

Example 5.1 Suppose that currents I_1 and I_2 are applied to the inner and outer superconductors in the coaxial superconducting transmission line shown in Fig. 5.5. Determine the magnetic flux density in each region.

Fig. 5.5 Coaxial superconducting transmission line

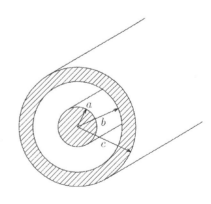

Solution 5.1 Current flows uniformly on the surface at $r = a$, so that the inside is completely shielded in the inner superconductor. Then, current also flows uniformly on the surface at $r = b$ to shield the outer superconductor. We assume a circle of

radius $r(a < r < b)$ with the center on the central axis, and apply Ampere's law on this circle. Since the magnetic flux density is zero on the circle, the total current inside the circle is zero. Thus, the current that flows on the surface at $r = b$ is $-I_1$. Hence, the current that flows on the outer surface of the outer superconductor at $r = c$ is $I_1 + I_2$.

Based on the above current distribution, the magnetic flux density is obtained to be

$$
\begin{aligned}
B &= 0; & 0 \leq r < a, \\
&= \frac{\mu_0 I_1}{2\pi r}; & a < r < b, \\
&= 0; & b < r < c, \\
&= \frac{\mu_0(I_1 + I_2)}{2\pi r}; & r > c.
\end{aligned}
$$

◇

Example 5.2 We apply current I to each wide slab superconductor in Fig. 5.6 in the backward normal direction to the page. Determine the currents on each surface and the magnetic flux density in each region. The width of the slabs is w.

Fig. 5.6 Two wide slab superconductors

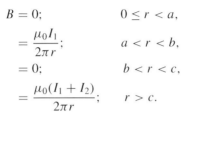

Solution 5.2 We denote by I_x the current that flows on the right surface of the left slab superconductor. Then, the current that flows on the left surface of the left superconductor is $I - I_x$, and the total current that flows on the right side of this superconductor is $I + I_x$. The two currents must be the same so that no magnetic flux density appears in this superconductor. Thus, we have

$$
I_x = 0.
$$

So, the current on the left surface of this superconductor is I. The current distribution in the right superconductor can also be obtained in a similar manner, and the currents on each surface are I, 0, 0, and I from the left. Thus, the magnetic flux density directed upward in each region is

$$B = \frac{\mu_0 I}{w}; \qquad \text{left vacuum,}$$

$$= 0; \qquad \text{two superconductors and space inbetween,}$$

$$= -\frac{\mu_0 I}{w}; \qquad \text{right vacuum.}$$

◇

5.2 Image Method

Suppose that current I flows at a distance a from infinitely wide superconductor surface, as shown in Fig. 5.7a. The current is induced on the superconductor surface to shield the inside against the magnetic flux produced by the applied current. It may seem to be difficult to determine the current distribution on the superconductor surface and the magnetic flux density in the space. There is a special and effective method to solve this problem, however. This is the **image method** introduced in Sect. 2.2, and we use the property that the magnetic flux density is parallel to the superconductor surface.

Here, we use Cartesian coordinates with the x-y plane ($z = 0$) on the superconductor surface. The y-axis ($x = 0$) is defined at the projection of the applied current on the surface. If there is no superconductor, we have only to place the opposite current $-I$ at the position symmetric with respect to the superconductor surface, as shown in Fig. 5.7b. The method of placing an opposite current at the position of the mirror image after virtual removal of the superconductor is also called the

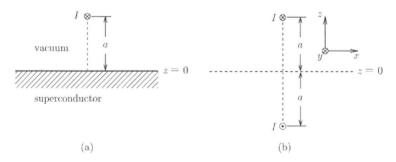

(a)   (b)

Fig. 5.7 a Straight current I placed at distance a from a wide superconductor surface and **b** image current $-I$ placed at the mirror position with respect to the superconductor surface after virtually removing the superconductor

image method. The virtual current at the image position is called the **image current**. The magnetic flux density produced by the given and image currents satisfies the condition that the magnetic flux density is parallel to the surface ($B_z = 0$) on the superconductor surface ($z = 0$), as shown in Fig. 5.8. So, the magnetic flux density produced by the two currents is correct in the vacuum area ($z > 0$). The magnetic flux density on the superconductor surface is directed towards the x-axis and its value is

$$B(x) = -\frac{\mu_0 I a}{\pi \left(x^2 + a^2\right)}. \tag{5.7}$$

From Eq. (5.3), the density of the superconducting current induced on the superconductor surface is

$$\tau(x) = \frac{B(x)}{\mu_0} = -\frac{I a}{\pi \left(x^2 + a^2\right)}. \tag{5.8}$$

Here, we shall determine the total current induced on the superconductor surface. It is given by

$$I' = \int_{-\infty}^{\infty} \tau(x) dx = -\frac{I a}{\pi} \int_{-\infty}^{\infty} \frac{dx}{x^2 + a^2} = -I. \tag{5.9}$$

This is equal to the image current. Next, we shall calculate the force between the given and induced currents. The force between the given current and the current induced in the region from x to $x + dx$, $\tau(x) dx$, is $dF = \mu_0 \tau(x) I dx / 2\pi \left(x^2 + a^2\right)^{1/2}$ in a unit length. The vertical component alone remains due to symmetry:

$$dF' = -\frac{a dF}{\left(x^2 + a^2\right)^{1/2}} = \frac{\mu_0 I^2 a^2 dx}{2\pi^2 \left(x^2 + a^2\right)^2}. \tag{5.10}$$

Fig. 5.8 The magnetic flux density on the superconductor surface produced by the given and image currents

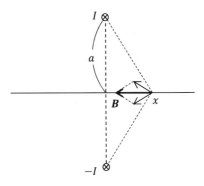

Hence, the total force in a unit length is

$$F = \frac{\mu_0 I^2 a^2}{2\pi^2} \int_{-\infty}^{\infty} \frac{\mathrm{d}x}{\left(x^2 + a^2\right)^2} = \frac{\mu_0 I^2}{2\pi^2 a} \int_{-\pi/2}^{\pi/2} \cos^2\theta\,\mathrm{d}\theta = \frac{\mu_0 I^2}{4\pi a}. \tag{5.11}$$

This is nothing else than the force between the given and image currents.

Example 5.3 The current induced on the superconductor surface was determined using the boundary condition on the superconductor surface in the above. Prove that the interior of the superconductor is completely shielded by the induced current given by Eq. (5.8).

Solution 5.3 The current induced on the superconductor surface also produces the magnetic flux density inside the superconductor. This magnetic flux density is symmetric with that produced in vacuum with respect to the surface. Hence, the magnetic flux density produced by the induced current in the superconductor is equal to that produced by a current $-I$ placed at the position of I. Namely, the total magnetic flux density is equal to that produced by I and $-I$ at the same place, i.e., the magnetic flux density when no current is given. Thus, the zero magnetic flux density in the superconductor can be proved.

◇

5.3 Coil and Inductance

Suppose that a surface S is surrounded by closed loop C. The surface integral of the magnetic flux density **B** on S is the **magnetic flux** that penetrates C (see Fig. 5.9):

Fig. 5.9 Magnetic flux penetrating surface S surrounded by closed loop C

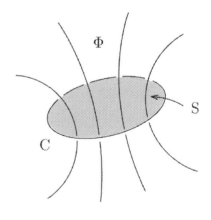

$$\Phi = \int_S \mathbf{B} \cdot d\mathbf{S}. \qquad (5.12)$$

This is determined by C but is independent on S. The unit of magnetic flux is $[\text{Tm}^2]$ and is newly defined as [Wb] (**weber**).

When a current is applied to a closed loop, the magnetic flux that is thus produced penetrates the closed loop. The resultant magnetic flux Φ is proportional to the current I and is expressed as

$$\Phi = LI. \qquad (5.13)$$

The proportional constant L is a positive quantity determined by the shape of the closed loop and is called the **inductance**. The unit of inductance is [Wb/A] and is newly defined as [H] (**henry**).

The electric device used to store the magnetic flux is the **coil**. It is usual to make a coil by winding wires in many turns to store a large amount of magnetic flux, as shown in Fig. 5.10. A coil with a cylindrical shape is called a **solenoid coil**. Here, we shall determine the inductance of a long solenoid coil with radius of the cross-section a, length l, and total number of turns N. When current I is applied, the magnetic flux is produced along the axial direction, as shown in Fig. 5.11. Here, we apply Ampere's law on a closed rectangle inside the coil as in (a) in the figure. Since there is no magnetic flux penetrating the rectangle, it can be said that the magnetic flux density inside the coil is uniform. The same result can be obtained outside the coil. Since it is not realistic that the magnetic flux density takes on a finite value even at a far distance, the magnetic flux density outside the coil is concluded to be zero. Then, we apply Ampere's law on a closed rectangle such as in (b) in Fig. 5.11. If we denote the magnetic flux density inside the coil by B, the left side of Eq. (4.17) is bB. Since the number of turns inside the rectangle is bN/l, the current penetrating it is bNI/l. Thus, we have $B = \mu_0 NI/l$. The magnetic flux penetrating one turn of the coil is

Fig. 5.10 Solenoid coil

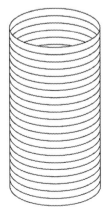

Fig. 5.11 Cross section of a
solenoid coil and rectangles
to which Ampere's law is
applied

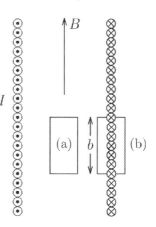

$$\phi = \pi a^2 B = \frac{\pi \mu_0 a^2 N I}{l} \qquad (5.14)$$

and the magnetic flux that penetrates the coil of N turns is

$$\Phi = N\phi = \frac{\pi \mu_0 a^2 N^2 I}{l}. \qquad (5.15)$$

Thus, we have the inductance:

$$L = \frac{\pi \mu_0 a^2 N^2}{l}. \qquad (5.16)$$

Suppose that there are two coils. When currents I_1 and I_2 flow in each coil, as shown in Fig. 5.12, the magnetic flux penetrating coil 1 is expressed as

$$\Phi_1 = L_{11} I_1 + L_{12} I_2, \qquad (5.17)$$

where the first term is a component produced by the current flowing in itself, and the second term is a component produced by the current flowing in the other coil. The magnetic flux penetrating coil 2 is similarly described as

$$\Phi_2 = L_{21} I_1 + L_{22} I_2. \qquad (5.18)$$

In the above L_{11} and L_{22} are **self-inductances**, and L_{12} and L_{21} are **mutual inductances**. For mutual inductances, the following relationship holds:

$$L_{12} = L_{21}. \qquad (5.19)$$

Fig. 5.12 Two coils in
which currents flow and
produced magnetic flux

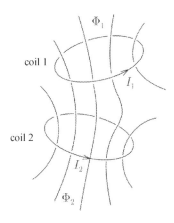

The self-inductance is positive and the mutual inductance has a positive or negative
value depending on the arrangement of the coils.

Example 5.4 Suppose a long coaxial superconducting transmission line, as shown
in Fig. 5.13. Determine the self-inductance of this transmission line.

Fig. 5.13 Coaxial
superconducting
transmission line

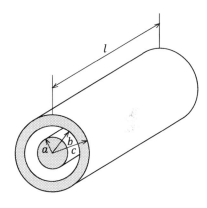

Solution 5.4 When currents I and $-I$ flow the inner and outer superconductors,
respectively, the magnetic flux appears only in the space, $a < r < b$, and its value is
$B = \mu_0 I / 2\pi r$. Thus, the total magnetic flux is determined to be

$$\Phi = \int dz \int_a^b \frac{\mu_0 I}{2\pi r} dr = \frac{\mu_0 I l}{2\pi} \log \frac{b}{a}.$$

Thus, we have the self-inductance:

$$L = \frac{\Phi}{I} = \frac{\mu_0 l}{2\pi} \log \frac{b}{a}.$$

◇

Example 5.5 Suppose two long solenoid coils with the same axis and the same length, as shown in Fig. 5.14. The radii of these coils are a_1 and a_2 ($a_1 > a_2$), the numbers of turns are N_1 and N_2, and the lengths are l. Determine the mutual inductance of these coils.

Fig. 5.14 Two solenoid coils with a common axis

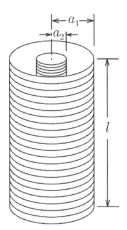

Solution 5.5 We apply current I_2 to the inner coil of radius a_2. The magnetic flux density that this coil produces in its interior is $B = \mu_0 N_2 I_2 / l$. Hence the magnetic flux that penetrates the outer coil is

$$\Phi_1 = \pi a_2^2 B N_1 = \frac{\pi \mu_0 a_2^2 N_1 N_2 I_2}{l}.$$

Thus, the mutual inductance is given by

$$L_{12} = \frac{\Phi_1}{I_2} = \frac{\pi \mu_0 a_2^2 N_1 N_2}{l}.$$

When we apply current I_1 to the outer coil of radius a_1, the magnetic flux density that this coil produces in its interior is $B = \mu_0 N_1 I_1 / l$. Hence, the magnetic flux that penetrates the inner coil is

$$\Phi_2 = \pi a_2^2 B N_2 = \frac{\pi \mu_0 a_2^2 N_1 N_2 I_1}{l},$$

and the mutual inductance is

$$L_{21} = \frac{\Phi_2}{I_1} = \frac{\pi\mu_0 a_2^2 N_1 N_2}{l}.$$

So, the relationship (5.19) holds.

◇

5.4 Magnetic Materials

When an electric field is applied to a dielectric material, electric dipole moments are induced due to the movement of polarization charges, and electric polarization takes place. If we apply an external magnetic flux density to a **magnetic material**, **magnetic moments** appear due to various mechanisms, and characteristic magnetic phenomena occur. This magnetic behavior is called **magnetization**, which is compared with electric polarization in dielectric materials. Electron spins and the orbital motion of electrons are known as the main origins of magnetic moments. It is well known that such magnetic properties can be equivalently expressed by a virtual closed current called **magnetizing current**, which cannot be transferred outside. On the other hand, the magnetization known for superconductors is caused by true currents. The magnetic moment in a unit volume is also called the magnetization, and is represented by M.

Suppose that current I flows on closed loop C, as shown in Fig. 5.15. If we represent by S the surface vector of this closed loop, which has a magnitude equal to the area of the loop and points along the direction of a screw when we rotate it in the current direction, the magnetic moment is given by

$$m = I S. \tag{5.20}$$

Suppose that a slab of a magnetic material is magnetized uniformly in the direction normal to its surface. When this slab is divided into small regions, as shown in Fig. 5.16a, the magnetic moment in each region can be expressed by the magnetizing current around it. These currents cancel out between adjacent planes, leaving only the magnetizing current flowing on the periphery of the slab, as shown in Fig. 5.16b.

Fig. 5.15 Magnetic moment
due to closed current

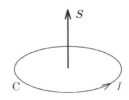

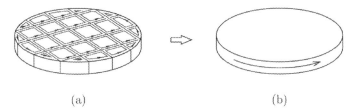

(a)   (b)

Fig. 5.16 a Aggregate of small closed currents and **b** resultant current on periphery

Here, we assume that a magnetic slab is uniformly magnetized by applying a parallel magnetic flux density, as shown in Fig. 5.17. We determine the magnetic moment of the magnetic slab. An equivalent magnetizing current flows on the surface of the magnetic material. We denote its surface density by τ_m. Then, the total current is $\tau_\mathrm{m}L$, and the total magnetic moment is $\tau_\mathrm{m}LS$. Since the volume of the magnetic slab is LS, from the definition, the magnetization is given by

$$M = \tau_\mathrm{m}. \tag{5.21}$$

Namely, the magnitude of the magnetization is equal to the surface density of the magnetizing current flowing on the surface. The unit of magnetization is [A/m].

The permanent magnet is known as one type of magnetic materials. There are N and S poles in a permanent magnet. The like poles of two magnets repel each other, while as attractive force works between different poles. So, the magnetic interaction between magnets may seem to be caused by particles much like electric charges. Such an imaginary charge is called a **magnetic charge**. In reality, magnetic charge does not exist. If a magnet is divided into small pieces, it is not possible to pick up one type of magnetic charge, and two types of magnetic poles exist, as shown in Fig. 5.18a. This situation can be explained by the equivalent magnetizing current, as shown in Fig. 5.18b.

Fig. 5.17 Magnetizing current flowing on the surface of uniformly magnetized slab

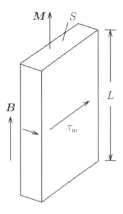

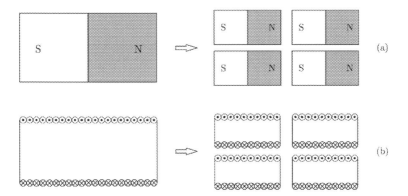

Fig. 5.18 **a** Divided permanent magnet and **b** divided equivalent magnetizing current

When a magnetic material exits, the magnetic flux density may be produced not only by true current, but also by magnetizing current. We denote by $\boldsymbol{B}_m$ the magnetic flux density produced by magnetizing current, then, we have

$$\boldsymbol{B}_m = \mu_0 \boldsymbol{M}. \tag{5.22}$$

If the magnetic flux density due to true current is given by $\mu_0 \boldsymbol{H}$, the total magnetic flux density is

$$\boldsymbol{B} = \mu_0 \boldsymbol{H} + \boldsymbol{B}_m = \mu_0 (\boldsymbol{H} + \boldsymbol{M}). \tag{5.23}$$

Then, $\boldsymbol{H}$ is given by

$$\boldsymbol{H} = \frac{1}{\mu_0} \boldsymbol{B} - \boldsymbol{M}. \tag{5.24}$$

The quantity given by this equation is called the **magnetic field** or **magnetic field strength**. Its unit is same as that of magnetization and is [A/m]. For the usual materials, except for ferromagnetic materials, etc., $\boldsymbol{v}$ is proportional to $\boldsymbol{H}$, and hence, $\boldsymbol{B}$ is also proportional to $\boldsymbol{H}$. Then, we can write $\boldsymbol{H}$ as

$$\boldsymbol{H} = \frac{\boldsymbol{B}}{\mu}, \tag{5.25}$$

and μ is called the **magnetic permeability**. If we write it as

$$\mu = \mu_0 \mu_r, \tag{5.26}$$

μ_r is called the **relative magnetic permeability**. When we write $\boldsymbol{M}$ as

Table 5.1 Magnetic susceptibility per kg/m^3 of non-magnetic materials

Material	$\chi (\times 10^{-3})$	Material	$\chi (\times 10^{-3})$
Diamond	− 0.49	Oxygen	106.2
Graphite	− 6 to − 7	Air	24.1
Gold	− 0.139	Nitrogen	− 0.43
Copper	− 0.086	Hydrogen	− 1.97
Zinc	− 0.157	Pure water	− 0.720
Germanium	− 0.12	Benzene	− 0.712
Aluminum	0.62	Quartz glass	− 0.5
Manganese	9.6	Alumina	− 0.34
Chromium	3.17	Iron dioxide	20.6

$$M = \chi H, \tag{5.27}$$

χ is called the **magnetic susceptibility**, and there is a relationship: $\mu_r = 1 + \chi$. The magnetic susceptibility of non-magnetic materials is listed in Table 5.1. Since $\mu_0 H$ is the magnetic flux density given by true current, **Ampere's law**, Eq. (4.17), is written as

$$\oint_C H \cdot ds = \oint_S i \cdot dS. \tag{5.28}$$

In other words, the current density in the right-hand side of Eq. (4.17) can include the magnetizing current density.

Example 5.6 Assume that a magnetic flux density B_0 is applied parallel to the wide surface of a magnetic material of magnetic permeability μ, as shown in Fig. 5.19. Determine the magnetic flux density and magnetization inside the magnetic material and the magnetizing current density on the surface.

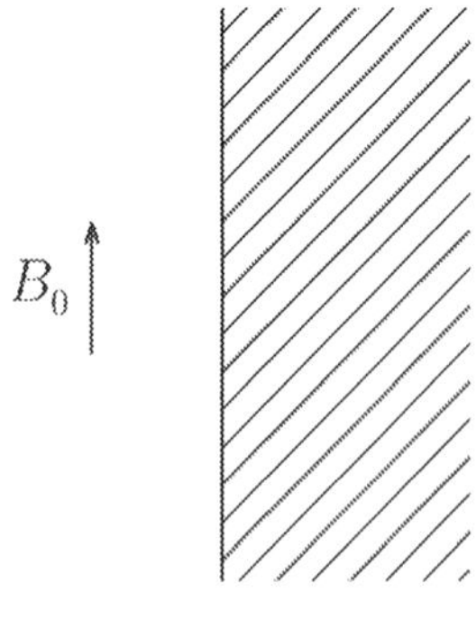

Fig. 5.19 Magnetic flux density B_0 applied parallel to the wide surface of a magnetic material

Solution 5.6 Suppose a closed rectangle C, as shown in Fig. 5.20a. We apply Ampere's law, Eq. (5.28), on C. Since true current does not flow on the surface of the magnetic material, the parallel component of the magnetic field strength is continuous between vacuum and the magnetic material. As the magnetic field strength in vacuum is B_0/μ_0, if we denote the magnetic flux density inside the magnetic material by B, the above continuity leads to $B_0/\mu_0 = B/\mu$. Thus, we have

$$B = \frac{\mu}{\mu_0} B_0. \tag{5.29}$$

Since $\mu > \mu_0$, the magnetic flux density is larger inside the magnetic material (see Fig. 5.20b). This is different from the smaller electric field strength inside a dielectric material due to the shielding by polarization charges, as argued in Example 2.5. Thus, the magnetization is determined to be

$$M = \left(\frac{1}{\mu_0} - \frac{1}{\mu} \right) B = \frac{(\mu - \mu_0)}{\mu_0^2} B_0.$$

From Eq. (5.21), the magnetizing current density on the surface is $\tau_m = M$. This can be directly derived from Eq. (4.17) (Note that i includes the magnetizing current density.).

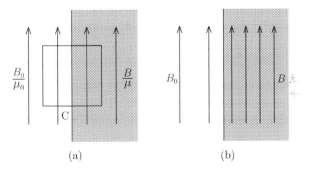

Fig. 5.20 Conditions of **a** the magnetic field and **b** the magnetic flux density at the surface

◇

From the result of Example 5.6, it is shown that the parallel component of the magnetic field strength is continuous at the interface between materials with different magnetic permeabilities, when there is no true current on the interface. If we denote by H_{1t} and H_{2t} the parallel components of the magnetic field strength in each material near the interface, we have

Fig. 5.21 a Closed surface
S at the interface between
materials with different
permeabilities and
b magnetic flux density on S

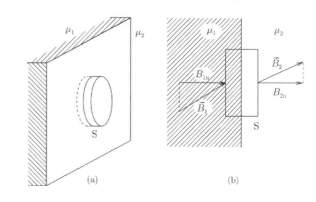

$$H_{1t} = H_{2t}. \tag{5.30}$$

When a true current flows on the interface, the continuity does not hold. Suppose that true current of surface density $\boldsymbol{\tau}$ flows on the interface. Then, we have

$$\boldsymbol{n} \times (\boldsymbol{H}_1 - \boldsymbol{H}_2) = \boldsymbol{\tau}, \tag{5.31}$$

where $\boldsymbol{H}_1$ and $\boldsymbol{H}_2$ are the magnetic field strengths in materials 1 and 2, respectively, and $\boldsymbol{n}$ is the unit vector normal to the interface and is directed from material 2 to material 1.

Here, we assume a small closed surface S that contains part of the interface between two materials with different magnetic permeabilities with two wider surfaces parallel to the interface, as shown in Fig. 5.21. The magnetic flux densities in the two magnetic materials are denoted by $\boldsymbol{B}_1$ and $\boldsymbol{B}_2$, respectively. If we apply Eq. (4.8) on S, we have

$$B_{1n} = B_{2n}, \tag{5.32}$$

where B_{1n} and B_{2n} are the normal components of $\boldsymbol{B}_1$ and $\boldsymbol{B}_2$, respectively. Thus, the normal component of the magnetic flux density is continuous on the interface. This condition is independent of the surface current. Equation (5.32) can be generally expressed as

$$\boldsymbol{n} \cdot (\boldsymbol{B}_1 - \boldsymbol{B}_2) = 0. \tag{5.33}$$

Here, we discuss the refraction of magnetic flux lines at the interface between different magnetic materials. Assume that magnetic flux density B_1 is applied in material 1 with magnetic permeability μ_1 in the direction of angle θ_1 from the normal direction to the interface, as shown in Fig. 5.22. The magnetic flux density B_2 and its angle θ_2 in the magnetic material with μ_2 are determined using the boundary conditions. From the continuity of the normal component of the magnetic flux density given by Eq. (5.32), we have

Fig. 5.22 Refraction of magnetic flux lines at interface

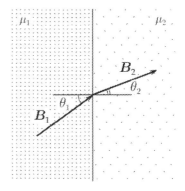

$$B_1\cos\theta_1 = B_2\cos\theta_2. \tag{5.34}$$

Equation (5.30) showing the continuity of the parallel component of the magnetic field strength gives

$$\frac{B_1}{\mu_1}\sin\theta_1 = \frac{B_2}{\mu_2}\sin\theta_2. \tag{5.35}$$

From Eqs. (5.34) and (5.35), we have

$$\frac{\tan\theta_1}{\tan\theta_2} = \frac{\mu_1}{\mu_2}. \tag{5.36}$$

This is the **law of refraction**. The value and angle of the magnetic flux density in material 2 are

$$B_2 = B_1\left[\left(\frac{\mu_2}{\mu_1}\right)^2\sin^2\theta_1 + \cos^2\theta_1\right]^{1/2}, \tag{5.37}$$

$$\theta_2 = \tan^{-1}\left(\frac{\mu_2}{\mu_1}\tan\theta_1\right). \tag{5.38}$$

Thus, the refraction of magnetic flux lines can be explained using the boundary conditions of magnetic flux density and magnetic field strength.

Example 5.7 When a magnetic rod of magnetic permeability μ is inserted into a long solenoid coil, as shown in Fig. 5.23, determine the self-inductance of the coil. The radius, length, and number of turns of the solenoid coil are a, l, and N, respectively, and the radius of the magnetic rod is $b(< a)$.

Fig. 5.23 Long solenoid
coil with magnetic rod of
magnetic permeability μ

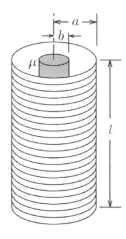

Solution 5.7 When we apply current I to the coil, the magnetic field strength inside
the coil is obtained using Eq. (5.28) to be $H = NI/l$. So, the magnetic flux density is
$B_1 = \mu_0 NI/l$ and $B_2 = \mu NI/l$ in vacuum and the magnetic material, respectively.
The magnetic flux that penetrates one turn of the coil is

$$\phi = \frac{\pi\left[\mu_0\left(a^2 - b^2\right) + \mu b^2\right]NI}{l},$$

and the magnetic flux that penetrates the coil is

$$\Phi = N\phi = \frac{\pi\left[\mu_0\left(a^2 - b^2\right) + \mu b^2\right]N^2 I}{l}.$$

Thus, the inductance is determined to be

$$L = \frac{\pi\left[\mu_0\left(a^2 - b^2\right) + \mu b^2\right]N^2}{l}.$$

When the inside of the coil is fully occupied by the magnetic material ($b = a$), the
inductance is $L = \pi\mu a^2 N^2/l$, which is μ/μ_0 times as large as the value given by
Eq. (5.16). The inductance can be increased by using magnetic materials.

$\diamondsuit$

Example 5.8 Two kinds of magnetic material occupy the space of a parallel-
plate superconducting transmission line, as shown in Fig. 5.24. Determine the self-
inductance of this transmission line, where its length along the direction normal to
the page is l.

Fig. 5.24 Cross-section of superconducting transmission line with two kinds of magnetic material

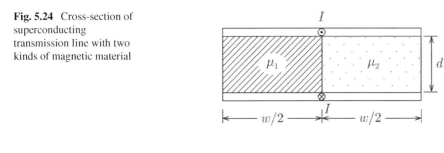

Solution 5.8 We apply current I to the superconducting transmission line. The magnetic flux density in the space between the two superconducting plates is parallel to the plates. So, the magnetic flux density is continuous across the interface from the boundary condition given by Eq. (5.32). This yields a difference of the magnetic field strength between the two magnetic materials, resulting in a difference in the current density flowing on the superconductor surface between the regions facing different magnetic materials. This is similar to the difference in the surface electric charge density on a conductor surface between the regions facing different dielectric materials in the capacitor treated in Example 2.7. Such a situation can be realized only for a superconductor in which Ohm's law does not hold. If we denote by B the magnetic flux density in both magnetic materials, the magnetic field strength in each magnetic material is

$$H_1 = \frac{B}{\mu_1}, \quad H_2 = \frac{B}{\mu_2}.$$

So, the current density on the superconducting plate in each region is

$$\tau_1 = \frac{B}{\mu_1}, \quad \tau_2 = \frac{B}{\mu_2},$$

and the total current is

$$I = \frac{w}{2}(\tau_1 + \tau_2) = \frac{wB(\mu_1 + \mu_2)}{2\mu_1\mu_2}.$$

Thus, we have

$$B = \frac{2\mu_1\mu_2 I}{w(\mu_1 + \mu_2)}.$$

The magnetic flux in the transmission line and the self-inductance are, respectively, given by

$$\Phi = Bld = \frac{2\mu_1\mu_2 ld I}{w(\mu_1 + \mu_2)},$$

and

$$L = \frac{2\mu_1\mu_2 ld}{w(\mu_1 + \mu_2)}.$$

◇

Exercises

5.1 Determine the current distribution in the outer superconductor and the magnetic flux density in each region, when current I is applied to the inner superconductor in the coaxial superconducting transmission line in Fig. 5.5.

5.2 We apply current I and $2I$ to the left and right wide superconducting slabs, respectively, in the backward normal direction to the page in Fig. 5.6. Determine the current on each surface.

5.3 Suppose that a magnetic flux density B_0 is applied normal to the wide surface of a magnetic material with magnetic permeability μ, as shown in Fig. 5.25. Determine the magnetic flux density, magnetic field strength, and magnetization in the magnetic material, and the magnetizing current density on the surface.

5.4 Determine the self-inductance of a parallel-plate transmission line shown in Fig. 5.26.

5.5 It is assumed that two circular coils of radius a are placed on the same axis with distance $2d$, as shown in Fig. 5.27. The z-axis is defined on the common

Fig. 5.25 Magnetic flux density B_0 applied normal to the surface of a magnetic material

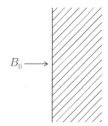

$B_0 \longrightarrow$

Fig. 5.26 Parallel-plate transmission line

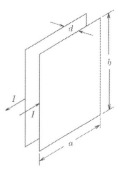

Fig. 5.27 Two circular coils
with the same radius placed
on the common axis

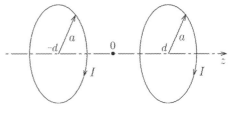

Fig. 5.28 Two parallel-wire
transmission lines

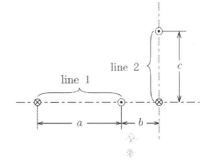

axis with the origin $z = 0$ on the position of the center of the two coils.
Determine the magnetic flux density on the common axis, when current I is
applied to the two coils. Use the solution of Exercise (5) in Chap. 4. To obtain
a uniform magnetic flux density over a fairly long distance on the common
axis, it is required to realize the condition $\partial^2 B/\partial z^2 = 0$ at $z = 0$. Show that
this condition is realized when $2d = a$. Such a pair of coils with the same
distance as the radius is called the **Helmholtz coil**.

5.6 Determine the mutual inductance in a unit length between the two parallel-
wire transmission lines in Fig. 5.28. We define the current directions as shown
in the figure.

5.7 Determine the mutual inductance between the parallel-wire transmission line
and a closed rectangle circuit, as shown in Fig. 5.29. These are placed on a
common plane, and the current direction shown by the arrows is defined as
positive.

5.8 Suppose a parallel-plate superconducting transmission line with the space
occupied by two magnetic materials with different magnetic permeabilities,
as shown in Fig. 5.30. Determine the inductance of this transmission line. Its
length is l.

Fig. 5.29 Parallel-wire
transmission line and a
closed rectangle circuit

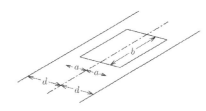

Fig. 5.30 Parallel-plate
superconducting
transmission line with two
magnetic materials

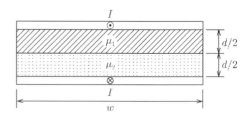

5.9   Determine the inductance of a unit length of a coaxial superconducting
      transmission line with two kinds of magnetic material in Fig. 5.31.

5.10 Determine the inductance of a unit length of a coaxial superconducting trans-
      mission line with two kinds of magnetic material in Fig. 5.32. The radii a and
      b are not so much different.

Fig. 5.31 Coaxial
superconducting
transmission line with two
kinds of magnetic material

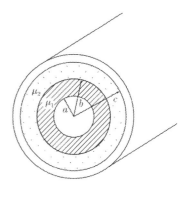

Fig. 5.32 Coaxial
superconducting
transmission line with two
kinds of magnetic material

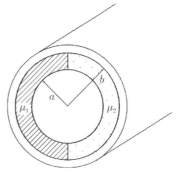

Reference

1. Matsushita T (2022) Electricity and magnetism, 2nd edn. Springer (point: Introduction of
   superconductivity)

Chapter 6
Time-Dependent Electromagnetic Phenomena

Abstract This chapter covers electromagnetic phenomena that change with time. One of them is expressed by Faraday's law on electromagnetic induction. We learn about magnetic energy as an equivalent electric energy by using this law. We also learn the principles behind the generation of electricity and transformers. Another one is the displacement current, which appears in a capacitor: Although true alternating current does not flow across the space between the electrodes, the displacement current flows instead of it. Maxwell's theory is completed with these new items. The skin effect and electromagnetic wave are treated as examples of these new items. We also learn the important law of reflection and Snell's law for refraction of electromagnetic wave.

6.1 Electromagnetic Induction

In a steady state in which electric and magnetic fields do not change with time, the electric field strength is given by the electric potential, and Eq. (1.17) holds for any closed loop C:

$$\oint_C E \cdot ds = 0. \tag{6.1}$$

When the magnetic flux Φ penetrating C changes with time, however, the electromotive force appears:

$$V_{em} = \oint_C E \cdot ds = -\frac{d\Phi}{dt}. \tag{6.2}$$

Such a phenomenon is called **electromagnetic induction**, and V_{em} is the **induced electromotive force**. Equation (6.2) is called **Faraday's law**. In the above, the direction of Φ is taken to be positive, when it points to the direction of motion of a screw under a rotation of a screw driver in the direction of the current in C (see Fig. 6.1). Thus, Eq. (6.2) shows that the electromotive force occurs to reduce a change in

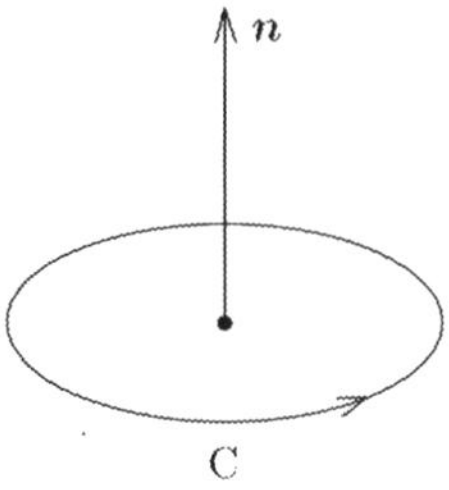

Fig. 6.1 Direction of the
magnetic flux along n is
positive when the current
flows along C

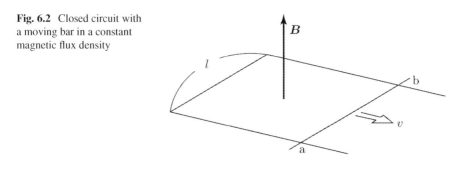

the magnetic flux penetrating the closed loop. The variation of the magnetic flux in
Eq. (6.2) includes the variation of the magnetic flux density and/or that of C with
time.

When a part ab of a closed circuit in a uniform magnetic flux density B moves
with speed v, as shown in Fig. 6.2, the electromotive force

$$V_{em} = -\frac{d\Phi}{dt} = -Bvl \tag{6.3}$$

appears in the circuit. Since other parts besides the bar ab does not move, the electromotive force is speculated to occur in ab. Hence, the **induced electric field** E has
a magnitude Bv and is directed from b to a. This is written in the form of a vector as

$$\boldsymbol{E} = \boldsymbol{v} \times \boldsymbol{B}, \tag{6.4}$$

which is known as **Fleming's right-hand rule**.

The equation that describes these phenomena, including the electromagnetic
induction, changes from Eq. (1.17) to

$$\oint_C \boldsymbol{E} \cdot d\boldsymbol{s} = -\frac{d}{dt} \int_S \boldsymbol{B} \cdot d\boldsymbol{S}. \tag{6.5}$$

Since this equation reduces to Eq. (1.17) when the magnetic flux density does not
change with time, this equation holds generally.

Fig. 6.2 Closed circuit with
a moving bar in a constant
magnetic flux density

Example 6.1 A rectangular coil of side lengths a and b and number of turns N rotates with angular velocity ω in a uniform magnetic flux density B, as shown in Fig. 6.3. The angle of the vector $\boldsymbol{n}$ normal to the coil surface from the applied magnetic flux density is denoted by $\theta = \omega t$. Determine the electromotive force induced in the coil. The direction shown by the arrows is positive.

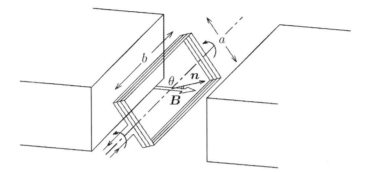

Fig. 6.3 Coil rotating in a uniform magnetic flux density

Solution 6.1 When current is applied to the coil in the direction shown by the arrows, the direction of the produced magnetic flux is the same as that of the applied magnetic flux density. Hence, the penetrating magnetic flux is positive. The magnetic flux penetrating one turn of the coil is $\phi = abB \cos \omega t$. Hence, the total magnetic flux penetrating the coil is $\Phi = N\phi = NabB \cos \omega t$. Thus, the electromotive force is determined to be

$$V_{em} = -\frac{d\Phi}{dt} = NabB\omega \sin \omega t.$$

This is the principle of **generation of alternating-current electricity**.

◇

When current I flowing in a coil of inductance L changes with time, the magnetic flux given by Eq. (5.13) changes and the electromotive force generated in the coil is given by

$$V_{em} = -L\frac{dI}{dt}, \tag{6.6}$$

and this is called **self-induction**. Suppose that there are two coils, as shown in Fig. 5.12. When the current I_1 flowing in coil 1 changes with time, the electromotive force due to the self-induction appears in coil 1 as shown above, and the electromotive force given by

$$V_2 = -L_{21}\frac{\mathrm{d}I_1}{\mathrm{d}t} \tag{6.7}$$

appears also in coil 2. This is called **mutual induction**.

Coil A of number of turns N_A and coil B of number of turns N_B are wound on a magnetic material of cross-sectional area S and large magnetic permeability μ, as shown in Fig. 6.4. Suppose that current $I_A = I \sin \omega t$ is applied to coil A and coil B is short-circuited. Then, we shall determine the current flowing in coil B. The magnetic flux in the magnetic material produced by the current in coil A is $\Phi = \mu S N_A I \sin \omega t$. Since the magnetic permeability is large, this magnetic flux penetrates coil B and the electromotive force in coil B is

$$V_B = V_{em} = -\mu S N_A N_B I \omega \cos \omega t. \tag{6.8}$$

We denote the current induced in coil B by I_B. The electromotive force that induces I_B is also written as $V_B = -L_B \mathrm{d}I_B/\mathrm{d}t$, using the self-inductance of coil B, $L_B = \mu S N_B^2$. Then, I_B is determined to be

$$I_B = \frac{N_A}{N_B} I \sin \omega t = \frac{N_A}{N_B} I_A. \tag{6.9}$$

On the other hand, the voltage of coil A is given by $V_A = -L_A \mathrm{d}I_A/\mathrm{d}t$, using the self-inductance of coil A, $L_A = \mu S N_A^2$. This shows that $V_B = (N_B/N_A)V_A$. Thus, we have

$$\frac{I_B}{I_A} = \frac{V_A}{V_B} = \frac{N_A}{N_B}. \tag{6.10}$$

This is the principle of the **transformer**.

Fig. 6.4   Coil A and coil B
wound on a magnetic
material

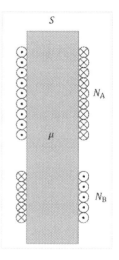

Example 6.2 Suppose that alternating current $I(t) = I_0 \sin \omega t$ flows in a straight wire, as shown in Fig. 6.5. Determine the electromotive force induced in the rectangular circuit with two parallel sides. The straight current and rectangular circuit are on the common plane. We define the electromotive force to be positive in the direction of ABCD.

Fig. 6.5 Straight wire with alternating current and rectangular circuit

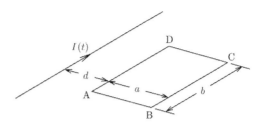

Solution 6.2 The direction of the magnetic flux due to current I that penetrates the circuit is opposite to that produced by the current flowing in the direction ABCD. The magnetic flux density at distance r from the current is $B(r) = \mu_0 I / 2\pi r$, and the magnetic flux that penetrates the circuit in the region from r to $r + dr$ is

$$d\Phi = -B(r)bdr = -\frac{\mu_0 I bdr}{2\pi r}.$$

Thus, the magnetic flux that penetrates the circuit is

$$\Phi = -\frac{\mu_0 I b}{2\pi} \int_d^{a+d} \frac{dr}{r} = -\frac{\mu_0 I b}{2\pi} \log \frac{a+d}{d},$$

and the induced electromotive force is determined to be

$$V_{em} = -\frac{d\Phi}{dt} = \frac{\mu_0 b}{2\pi} \cdot \frac{dI(t)}{dt} \log \frac{a+d}{d} = \frac{\mu_0 b I_0 \omega}{2\pi} \log \frac{a+d}{d} \cos \omega t.$$

$\diamond$

Example 6.3 Suppose that a rectangular circuit is moving away with velocity v from a straight current I, as shown in Fig. 6.6. Determine the electromotive force induced in the circuit using Eq. (6.4). The straight current and the rectangular circuit are placed on a common plane, and the electromotive force is defined to be positive in the direction of ABCD.

Fig. 6.6 Straight current I
and rectangular circuit
moving away from the
current

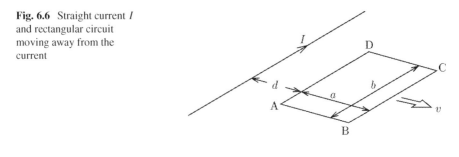

Solution 6.3 The electric field induced in sides AB and CD is perpendicular to each side, and hence, there is no contribution from these sides to the electromotive force. The induced electric field is $\mu_0 I v/[2\pi(d+a)]$ and is in the same direction to that of integration for side BC. So, the contribution to the electromotive force from this side is $\mu_0 I bv/[2\pi(d+a)]$. On side DA, the electric field is $\mu_0 I v/(2\pi d)$ and is directed opposite to the direction of integration. The contribution from this side is $-\mu_0 I bv/(2\pi d)$. As a result, the electromotive force induced in the circuit is determined to be

$$V_{em} = \frac{\mu_0 I bv}{2\pi(d+a)} - \frac{\mu_0 I bv}{2\pi d} = -\frac{\mu_0 I abv}{2\pi d(d+a)}.$$

◇

6.2   Magnetic Energy

We treat the process in which current is applied to a coil of inductance L. When the value of the current is I', the magnetic flux that penetrates the coil is $\Phi' = LI'$ and the induced electromotive force is

$$V_{em} = -L\frac{dI'}{dt} \tag{6.11}$$

and works to prevent variation of the current with time. Thus, it is necessary for the electric power source to supply the voltage

$$V = L\frac{dI'}{dt}, \tag{6.12}$$

to continue increasing the current. Equation (6.12) is the voltage drop over the coil. The electric power supplied by the power source is

$$P = VI' = LI'\frac{dI'}{dt}. \tag{6.13}$$

Thus, the energy supplied by the power source until the current reaches I is

$$U_m = \int LI'\frac{dI'}{dt}dt = L\int_0^I I'dI' = \frac{1}{2}LI^2. \tag{6.14}$$

This is called the **magnetic energy**. Using the relationship in Eq. (5.13), the magnetic energy is also expressed as

$$U_m = \frac{1}{2}LI^2 = \frac{1}{2}\Phi I = \frac{1}{2L}\Phi^2. \tag{6.15}$$

When current I is applied to a solenoid coil of radius a, length l, and number of turns N, as shown in Fig. 5.10, the magnetic energy is given by Eqs. (5.16) and (6.15). When this is expressed using the magnetic flux density, $B = \mu_0 NI/l$, we have

$$U_m = \frac{1}{2\mu_0}B^2\pi a^2 l. \tag{6.16}$$

In the above $\pi a^2 l$ is the volume of the space in which the magnetic flux is concentrated with a uniform density. So, we can regard

$$u_m = \frac{1}{2\mu_0}B^2 \tag{6.17}$$

as the magnetic energy in a unit volume. This is called the **magnetic energy density**. When the space of the solenoid coil is occupied by a magnetic material of magnetic permeability μ, we can replace μ_0 in Eq. (6.17) by μ.

Example 6.4 Determine the magnetic energy of the parallel-plate superconducting transmission line in Example 5.8, when current I is applied to it. Determine also the self-inductance with the obtained result.

Solution 6.4 The magnetic flux density in magnetic materials 1 and 2 is the same and is given by $B = 2\mu_1\mu_2 I/[w(\mu_1 + \mu_2)]$. Thus, the magnetic energy density in each magnetic material is

$$u_{m1} = \frac{2\mu_1\mu_2^2 I^2}{w^2(\mu_1 + \mu_2)^2}, \quad u_{m2} = \frac{2\mu_1^2\mu_2 I^2}{w^2(\mu_1 + \mu_2)^2}.$$

Then, the total energy is

$$U_\mathrm{m} = \frac{wld}{2}(u_\mathrm{m1} + u_\mathrm{m2}) = \frac{\mu_1 \mu_2 ld I^2}{w(\mu_1 + \mu_2)},$$

and the self-inductance is determined to be

$$L = \frac{2U_\mathrm{m}}{I^2} = \frac{2\mu_1 \mu_2 ld}{w(\mu_1 + \mu_2)}.$$

This agrees with the result in Example 5.8.

$\diamond$

6.3   Displacement Current

We have treated steady currents which do no change with time and magnetic phenomena caused by steady currents. Here, we treat the case in which current changes with time. Suppose a closed line C that forms the boundary of two different hemispherical surfaces S_1 and S_2, as shown in Fig. 6.7a, b. Ampere's law given by Eq. (5.28) holds for each of S_1 and S_2. If we sum up Ampere's law for each case, the sum of the circular integral on C is naturally zero, since the direction of the integral is opposite between (a) and (b). The corresponding sum of the surface integral is the surface integral on the closed surface formed by S_1 and S_2, which is now denoted by S_{12} (see Fig. 6.7c), and we have

$$\int_{S_{12}} i \cdot dS = 0, \tag{6.18}$$

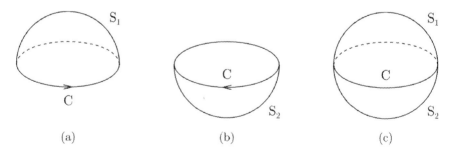

Fig. 6.7   Closed loop C and two different surfaces **a** S_1 and **b** S_2 on it, and **c** closed surface S_{12} composed of S_1 and S_2

and Eq. (3.5) for a steady state is derived. This means, however, that Ampere's law contradicts Eq. (3.4) for non-steady states. If Eq. (2.27) is substituted in Eq. (3.4), we have

$$\int_{S_{12}} \left(i + \frac{\partial D}{\partial t} \right) \cdot dS = 0. \tag{6.19}$$

Thus, if we assume

$$\oint_C H \cdot ds = \int_S \left(i + \frac{\partial D}{\partial t} \right) \cdot dS \tag{6.20}$$

instead of Ampere's law, there is no contradiction, where S is a surface on C. The second term, $\partial D / \partial t$, in this equation is called the **displacement current** and is different from the true current, i.e., a flow of true charges. This equation leads to Ampere's law in the steady state. Hence, there is no problem. Equation (6.20) is called the **generalized Ampere's law**.

Here, we show an example of the displacement current in a capacitor. When current I flows into a capacitor, as shown in Fig. 6.8a, electric charge Q on the electrode changes. We assume a closed loop C around a wire through which the current flows and surface S_1 as in this figure. Then, the displacement current is zero, and only current density i exists on the right side of Eq. (6.20). Thus, the right side of the equation leads to the current I.

Next, we assume another surface S_2 that does not contain the wire, as shown in Fig. 6.8b. On this surface, i is zero and the right side of Eq. (6.20) is

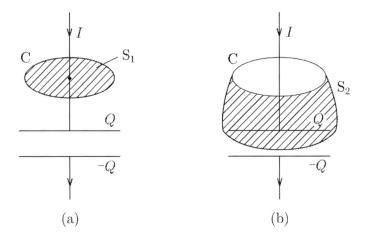

(a)   (b)

Fig. 6.8 Closed line C around a current-carrying wire and surface surrounded by C: **a** surface that includes the wire and **b** surface that does not include the wire

$$\frac{\partial}{\partial t}\int_{S_2} \boldsymbol{D}\cdot \mathrm{d}\boldsymbol{S} = \frac{\mathrm{d}Q}{\mathrm{d}t}. \tag{6.21}$$

Comparing the above two cases, we have

$$I = \frac{\mathrm{d}Q}{\mathrm{d}t}. \tag{6.22}$$

Thus, no contradiction results from Eq. (6.20).

Example 6.5 We apply alternating current $I(t) = I_0 \sin \omega t$ to a capacitor with circular electrodes of radius a separated by d, as shown in Fig. 6.9. Determine the displacement current in the space between the electrodes.

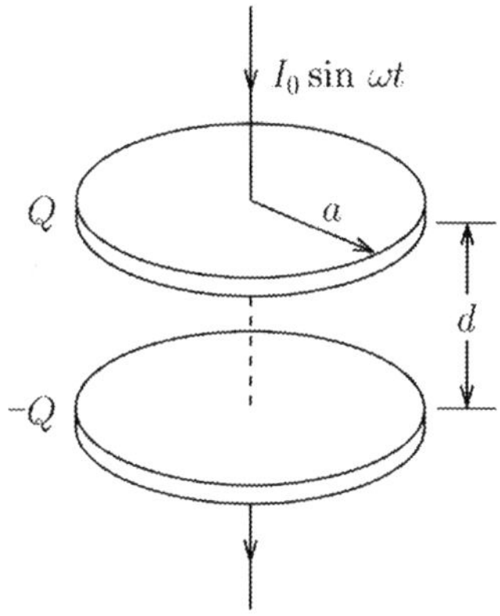

Fig. 6.9 Alternating current flowing through circular parallel-plate capacitor

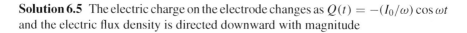

Solution 6.5 The electric charge on the electrode changes as $Q(t) = -(I_0/\omega)\cos \omega t$ and the electric flux density is directed downward with magnitude

$$D(t) = \frac{Q(t)}{\pi a^2} = -\frac{I_0}{\pi a^2 \omega}\cos \omega t.$$

Hence, the displacement current is

$$\frac{\partial D(t)}{\partial t} = \frac{I_0}{\pi a^2}\sin \omega t.$$

This is similar to a virtual situation in which current of the same density flows uniformly in the space between the electrodes, suggesting the continuity of current. In fact, if the space is occupied by a material with electric resistivity sufficiently higher than that of the electrodes, a true current of the same magnitude flows. ◇

6.4 Maxwell's Equations

The important equations that we have learned are

$$\oint_C E \cdot ds = -\frac{d}{dt} \int_S B \cdot dS \tag{6.23}$$

for the electromagnetic induction and

$$\oint_C H \cdot ds = \int_S \left(i + \frac{\partial D}{\partial t} \right) \cdot dS \tag{6.24}$$

for the current. In the above, S is a surface surrounded by a closed loop C. These equations also hold in a steady state in which the corresponding terms on the right side are constant. Other equations are

$$\int_{S'} D \cdot dS = \int_V \rho dV \tag{6.25}$$

for the electric flux and

$$\int_{S'} B \cdot dS = 0 \tag{6.26}$$

for the magnetic flux, where S' is a closed surface on a region V. These are called **Maxwell's equations**. The material relationships are given by

$$D = \epsilon E, \tag{6.27}$$

$$H = \frac{1}{\mu} B, \tag{6.28}$$

$$i = \sigma_c E. \tag{6.29}$$

Equations (6.23)–(6.26) are expressed as integral equations. Their differential equations are shown in Appendix A.2.

6.5 Skin Effect

Various dynamic electromagnetic phenomena can be solved using the above Maxwell's equations. Here, we show the skin effect as an example of dynamic phenomena. Suppose that an alternating electric field of amplitude E_0 and angular frequency ω is applied parallel to the surface of a semi-infinite material that occupies $x \geq 0$, in the direction of the y-axis. We investigate how the electric field penetrates into the material. We denote the dielectric constant, magnetic permeability, and electric conductivity by ϵ, μ, and σ_c, respectively.

Since the displacement current can be neglected in a usual conductive material, Eq. (6.24) leads to

$$\frac{1}{\mu} \oint_C \boldsymbol{B} \cdot d\boldsymbol{s} = \sigma_c \int_S \boldsymbol{E} \cdot d\boldsymbol{S}. \tag{6.30}$$

We can assume that the electric charge density ρ is zero. Then, Eq. (6.25) is reduced to

$$\int_{S'} \boldsymbol{E} \cdot d\boldsymbol{S} = 0. \tag{6.31}$$

Equation (6.23) holds.

Since the external electric field is applied in the direction of the y-axis, we can assume that the interior electric field has only the y-component. In addition, we can also assume that the quantities are uniform in the wide y-z plane, which allows us to presume that they vary along only the x-axis. It is found from Eq. (6.23) that the magnetic flux density has a component normal to the electric field. Equation (6.26) shows that it is the z-component as follows: Suppose a rectangular parallelepiped with each surface normal to the x, y, or z-axis for S′ in Eq. (6.26). If the magnetic flux density has an x-component, its value must be a constant up to infinity, which is not realistic.

Suppose a rectangle C in the regions from x to $x + \Delta x$ and from y to $y + \Delta y$ in Eq. (6.23), as shown in Fig. 6.10. If Δx and Δy are sufficiently small, the left side becomes

$$[E(x + \Delta x) - E(x)]\Delta y \simeq \frac{\partial E}{\partial x} \Delta x \Delta y. \tag{6.32}$$

Since the right side reduces to $-(\partial B/\partial t)\Delta x \Delta y$, we have

$$\frac{\partial E}{\partial x} = -\frac{\partial B}{\partial t}. \tag{6.33}$$

Fig. 6.10 Closed loop C in
Eq. (6.23)

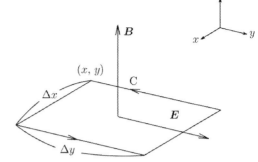

Suppose a rectangle C in the regions from x to $x + \Delta x$ and from z to $z + \Delta z$ in Eq. (6.30), as shown in Fig. 6.11. If Δx and Δz are sufficiently small, the left side becomes

$$\frac{1}{\mu}[-B(x + \Delta x) + B(x)]\Delta z \simeq -\frac{1}{\mu} \cdot \frac{\partial B}{\partial x}\Delta x \Delta z. \qquad (6.34)$$

Since the right side becomes $\sigma_c E \Delta x \Delta z$, we have

$$\frac{\partial B}{\partial x} = -\mu \sigma_c E. \qquad (6.35)$$

From Eqs. (6.33) and (6.35), the diffusion equation is derived for E:

$$\frac{\partial^2 E}{\partial x^2} = \mu \sigma_c \frac{\partial E}{\partial t}. \qquad (6.36)$$

The same equation is obtained for B. If we assume that E varies with time as $e^{i\omega t}$ with angular frequency ω, we can replace the time differentiation, $\partial/\partial t$, by multiplication by $i\omega$. Thus, Eq. (6.36) leads to

Fig. 6.11 Closed loop C in
Eq. (6.30)

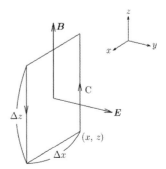

$$\frac{\partial^2 E}{\partial x^2} - i\omega\mu\sigma_c E = 0. \tag{6.37}$$

Assume a solution of type $E(x) \sim e^{\alpha x}$. Substituting this into Eq. (6.37), we have $\alpha^2 = i\omega\mu\sigma_c$. That is,

$$\alpha = \pm(1 + i)\left(\frac{\omega\mu\sigma_c}{2}\right)^{1/2}. \tag{6.38}$$

From the condition that the electric field strength must be finite in the limit $x \to \infty$, α with the negative real part is the solution. Using the boundary condition, $E(x = 0) = E_0$, the electric field strength is determined to be

$$E(x, t) = E_0 e^{-x/\delta} \exp\left[i\left(\omega t - \frac{x}{\delta}\right)\right], \tag{6.39}$$

where

$$\delta = \left(\frac{2}{\omega\mu\sigma_c}\right)^{1/2} \tag{6.40}$$

is a distance called the **skin depth**. Taking the real part, we have

$$E(x, t) = E_0 e^{-x/\delta} \cos\left(\omega t - \frac{x}{\delta}\right). \tag{6.41}$$

Figure 6.12a shows the spatial variation in the electric field strength. The electric field strength propagates along the depth from the surface while decaying. Thus, the depth of the electric field strength is roughly given by δ. This is the reason why this

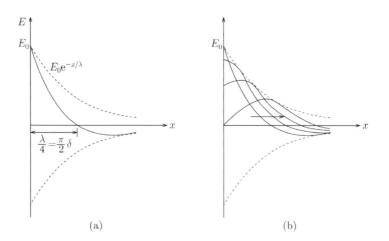

(a)   (b)

Fig. 6.12 a Spatial variation ($t = 0$) and **b** time variation in electric field strength

phenomenon is called the **skin effect**. For larger ω and/or σ_c, the shielding current density is higher, resulting in a shorter δ. The position of a plane on which the phase of the propagating wave is constant is given by the condition

$$\omega t - \frac{x}{\delta} = \text{const.} \tag{6.42}$$

Thus, the speed of propagation is $dx/dt = \omega\delta$ and the wave length is $2\pi\delta$.

Substituting Eq. (6.39) into Eq. (6.35), the magnetic flux density is determined to be

$$
\begin{aligned}
B(x, t) &= \frac{1-i}{2} \mu \sigma_c \delta E_0\, e^{-x/\delta} \exp\left[i\left(\omega t - \frac{x}{\delta}\right)\right] \\
&= E_0 \left(\frac{\mu\sigma_c}{\omega}\right)^{1/2} e^{-x/\delta} \exp\left[i\left(\omega t - \frac{x}{\delta} - \frac{\pi}{4}\right)\right],
\end{aligned} \tag{6.43}
$$

where we used the relation $(\partial/\partial x)^{-1} = -(1-i)\delta/2 = -\exp(-\pi i/4)\pi\delta/2^{1/2}$. Taking the real part, we have

$$B(x, t) = E_0 \left(\frac{\mu\sigma_c}{\omega}\right)^{1/2} e^{-x/\delta} \cos\left(\omega t - \frac{x}{\delta} - \frac{\pi}{4}\right). \tag{6.44}$$

Thus, the phase of the magnetic flux density lags behind that of the electric field by $\pi/4$.

6.6 Electromagnetic Waves

Here, we shall treat electromagnetic waves in vacuum. Suppose that the electric field strength has only a y-component. There is no electric charge ($\rho = 0$) in vacuum, and Eq. (6.25) is reduced to Eq. (6.31). Thus, it can be shown that the electric field strength does not depend on y. Namely, if we assume a rectangular parallelepiped with each surface normal to the x, y, or z-axis for S′, there are contributions only from the surface normal to the y-axis to the integral, but these are zero. It is shown by Eq. (6.23) that the magnetic flux density has a component normal to the electric field. We assume that the magnetic flux density has a z-component. We suppose C as shown in Fig. 6.11, and then, Eq. (6.23) leads to Eq. (6.33).

The current density i is also zero in vacuum, and Eq. (6.24) leads to

$$\frac{1}{\mu_0} \oint_C \boldsymbol{B} \cdot d\boldsymbol{s} = \epsilon_0 \int_S \frac{\partial \boldsymbol{E}}{\partial t} \cdot d\boldsymbol{S}. \tag{6.45}$$

It can be shown that the magnetic flux density does not depend on z from Eq. (6.26). If we assume again closed loop C in Eq. (6.45), as shown in Fig. 6.11, the left side is

$$\frac{1}{\mu_0}[-B(x + \Delta x) + B(x)]\Delta z \simeq -\frac{1}{\mu_0} \cdot \frac{\partial B}{\partial x}\Delta x \Delta z. \tag{6.46}$$

The right side becomes $\epsilon_0 (\partial E / \partial t)\Delta x \Delta z$, and we have

$$\frac{\partial B}{\partial x} = -\epsilon_0 \mu_0 \frac{\partial E}{\partial t}. \tag{6.47}$$

From Eqs. (6.33) and (6.47), the following equation is derived:

$$\frac{\partial^2 E}{\partial x^2} = \epsilon_0 \mu_0 \frac{\partial^2 E}{\partial t^2}. \tag{6.48}$$

We have the same equation for the magnetic flux density. An equation of this type is called a **wave equation**. If we assume that E varies with time as $e^{i\omega t}$ with angular frequency ω, we can replace the time differentiation, $\partial / \partial t$, by multiplication by $i\omega$. Thus, Eq. (6.48) leads to

$$\frac{\partial^2 E}{\partial x^2} + \epsilon_0 \mu_0 \omega^2 E = 0. \tag{6.49}$$

This is easily solved as

$$E(x, t) = E_1 \exp[i(\omega t + kx)] + E_2 \exp[i(\omega t - kx)], \tag{6.50}$$

where E_1 and E_2 are constants determined by the boundary conditions and

$$k = (\epsilon_0 \mu_0)^{1/2}\omega \tag{6.51}$$

is the wave number. In the first term in Eq. (6.50),

$$\omega t + kx = \text{const.} \tag{6.52}$$

represents the time-variation of the position of the same phase. This yields

$$\frac{dx}{dt} = -\frac{\omega}{k} = -\frac{1}{(\epsilon_0 \mu_0)^{1/2}} \equiv -c, \tag{6.53}$$

and the first term in Eq. (6.50) is the component of **electromagnetic wave** propagating in the direction of the negative x-axis with speed c. The second term is similarly the component of the electromagnetic wave propagating in the direction of the positive x-axis. Thus, c given by Eq. (6.53) is the **light speed** in vacuum. Since the same phase in Eq. (6.52) is obtained in a flat plane, such an electromagnetic wave is called a **planar electromagnetic wave**. The wave length is given by

$$\lambda = \frac{2\pi}{k} = \frac{2\pi}{(\epsilon_0\mu_0)^{1/2}\omega} = \frac{2\pi c}{\omega}. \tag{6.54}$$

Substitution of Eq. (6.50) into Eq. (6.33) yields the solution for the magnetic flux density:

$$B(x,t) = -B_1 \exp[i(\omega t + kx)] + B_2 \exp[i(\omega t - kx)]$$
$$= -\frac{E_1}{c} \exp[i(\omega t + kx)] + \frac{E_2}{c} \exp[i(\omega t - kx)]. \tag{6.55}$$

The first and second terms correspond to the first and second terms in Eq. (6.50), respectively.

Using the wave number vector k, the planar electromagnetic wave propagating in this direction is generally represented as

$$\exp[i(\omega t - k \cdot r)]. \tag{6.56}$$

Here, we treat the energy of electromagnetic waves. Assume a planar electromagnetic wave propagating in the direction of the positive x-axis. The real part of the electric field strength in the second term in Eq. (6.50) is

$$E = E_2 \cos(\omega t - kx). \tag{6.57}$$

Hence, the electric energy density is

$$u_e = \frac{1}{2}\epsilon_0 E_2^2 \cos^2(\omega t - kx). \tag{6.58}$$

The corresponding magnetic flux density is from Eq. (6.55) and given by

$$B = B_2 \cos(\omega t - kx) = \frac{E_2}{c} \cos(\omega t - kx), \tag{6.59}$$

and the magnetic energy density is

$$u_m = \frac{1}{2\mu_0}\left(\frac{E_2}{c}\right)^2 \cos^2(\omega t - kx) = \frac{1}{2}\epsilon_0 E_2^2 \cos^2(\omega t - kx), \tag{6.60}$$

which is the same as the electric energy density. Thus, the total energy density of the electromagnetic wave is given by

$$u = u_e + u_m = \epsilon_0 E_2^2 \cos^2(\omega t - kx) = \frac{B_2^2}{\mu_0} \cos^2(\omega t - kx). \tag{6.61}$$

As shown above, the electric field strength and magnetic flux density are perpendicular to each other, and these are perpendicular to the propagation direction. So, a planar electromagnetic wave is a transverse wave. The ratio of these quantities is

$$\frac{E_1}{B_1} = \frac{E_2}{B_2} = c. \tag{6.62}$$

The electromagnetic wave is a wave in which the electric field strength and magnetic flux density induce each other, and since there is no energy dissipation because of no electric charge, it propagates without decaying. Electromagnetic waves exist also in dielectric materials, and their propagation speed is

$$c = \frac{1}{(\epsilon\mu)^{1/2}}. \tag{6.63}$$

The magnetic field strength has been commonly used instead of the magnetic flux density to describe electromagnetic waves. In this case, Eq. (6.55) is written as

$$H(x,t) = -H_1 \exp[i(\omega t + kx)] + H_2 \exp[i(\omega t - kx)]. \tag{6.64}$$

The ratio of amplitudes

$$\frac{E_1}{H_1} = \frac{E_2}{H_2} = \mu_0 c = \left(\frac{\mu_0}{\epsilon_0}\right)^{1/2} \equiv Z \tag{6.65}$$

is the **wave impedance**, and its unit is [Ω].

Example 6.6 When true current and displacement current coexist in some material, derive the equation on the electric field strength, which was given by Eqs. (6.36) or (6.48) for each case. The dielectric constant, magnetic permeability, and electric conductivity are ϵ, μ, and σ_c, respectively, and it is assumed that there is no electric charge.

Solution 6.6 The equation for the magnetic flux density associated with the current changes from Eqs. (6.35) and (6.47), and now is given by

$$\frac{\partial B}{\partial x} = -\epsilon\mu\frac{\partial E}{\partial t} - \mu\sigma_c E.$$

The equation for the induced electric field is the same as Eq. (6.33). Eliminating B from these equations yields

$$\frac{\partial^2 E}{\partial x^2} = \epsilon\mu\frac{\partial^2 E}{\partial t^2} + \mu\sigma_c\frac{\partial E}{\partial t}.$$

This equation is called the **telegraphic equation**.

◇

Example 6.7 Determine the real part of the displacement current in the electromagnetic wave given by Eqs. (6.50) and (6.55).

Solution 6.7 From Eq. (6.50), the real part of the electric field strength is

$$E(x, t) = E_1 \cos(\omega t + kx) + E_2 \cos(\omega t - kx).$$

Hence, the displacement current is determined to be

$$\frac{\partial}{\partial t} D(x, t) = \epsilon_0 \frac{\partial}{\partial t} E(x, t) = \epsilon_0 \omega E_1 \sin(\omega t + kx) + \epsilon_0 \omega E_2 \sin(\omega t - kx).$$

As can be seen from Eq. (6.47), the same result can be obtained from the magnetic flux density given by Eq. (6.55).

◇

Here, we treat the reflection and refraction of planar electromagnetic wave. Suppose a planar interface $z = 0$ between media 1 and 2 with dielectric constants ϵ_1 and ϵ_2 and magnetic permeabilities μ_1 and μ_2, respectively, as shown in Fig. 6.13a, and a planar electromagnetic wave propagates from medium 1 to the interface. The plane formed by the propagation direction and the direction (z-axis) normal to the interface is called the plane of incidence. We define the x-axis as the line at which the plane of incidence and the interface meet and the y-axis on the interface in such a way that it is normal to both the x- and z-axes.

In this case the incident and reflected waves are in medium 1, and the transmitted wave is in medium 2. We denote by k, k'', and k' the wavenumbers of the incident,

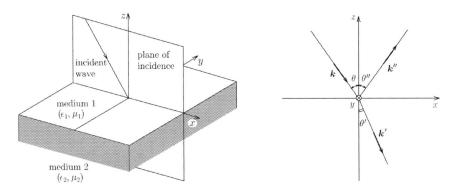

Fig. 6.13 Definition of **a** axes on the interface and **b** angles of waves

reflected, and transmitted waves, respectively. Each wave propagates in the direction of its wavenumber vector and these wavenumber vectors lie on the plane of incident (x-z plane). The angles of the wave numbers, k, k'', and k', from the z-axis are denoted by θ, θ'', and θ', respectively, as shown in Fig. 6.13b. The variation of each wave with time is the same, and the phase is given by ωt. The incident, refracted, and transmitted waves are expressed as

$$\exp[\mathrm{i}(\omega t - k \cdot r)], \quad \exp[\mathrm{i}(\omega t - k'' \cdot r)], \quad \exp[\mathrm{i}(\omega t - k' \cdot r)], \tag{6.66}$$

where r is the position vector.

The electric field strength E and the magnetic flux density B are perpendicular to each other, and both lie on a plane normal to the propagation direction. For example, E and B are normal to k for the incident wave. Here, we treat the boundary conditions to be satisfied for E and B. These are given by Eqs. (2.30), (2.32), (5.31), and (5.33). When there is no electric charge and no current on the interface, the parallel components of the electric field strength and magnetic field strength are continuous, and the normal components of the magnetic flux density and electric flux density are continuous across the interface. That is,

$$n \times (E_1 - E_2) = 0, \tag{6.67}$$

$$n \cdot (\epsilon_1 E_1 - \epsilon_2 E_2) = 0, \tag{6.68}$$

$$n \times \left(\frac{B_1}{\mu_1} - \frac{B_2}{\mu_2} \right) = 0, \tag{6.69}$$

$$n \cdot (B_1 - B_2) = 0, \tag{6.70}$$

where n is the unit vector normal to the interface, and the subscripts 1 and 2 represent the quantities in media 1 and 2, respectively.

Considering the orthogonality between the electric field strength and the magnetic flux density, the incident wave is given by

$$E = E_0 \exp[\mathrm{i}(\omega t - k \cdot r)], \tag{6.71a}$$

$$B = \frac{k}{k} \times \frac{E_0}{c_1} \exp[\mathrm{i}(\omega t - k \cdot r)], \tag{6.71b}$$

where $k = |k|$ and c_1 is the light speed in medium 1. The magnetic flux density B is generally normal to both E and k, and its magnitude is equal to the magnitude of E divided by the corresponding light speed. The reflected wave is similarly given by

$$E'' = E_0'' \exp[\mathrm{i}(\omega t - k'' \cdot r)], \tag{6.72a}$$

$$\boldsymbol{B}'' = \frac{\boldsymbol{k}''}{k''} \times \frac{E_0''}{c_1} \exp\left[i(\omega t - \boldsymbol{k}'' \cdot \boldsymbol{r})\right], \tag{6.72b}$$

and the transmitted wave is given by

$$\boldsymbol{E}' = E_0' \exp\left[i(\omega t - \boldsymbol{k}' \cdot \boldsymbol{r})\right], \tag{6.73a}$$

$$\boldsymbol{B}' = \frac{\boldsymbol{k}'}{k'} \times \frac{E_0'}{c_2} \exp\left[i(\omega t - \boldsymbol{k}' \cdot \boldsymbol{r})\right]. \tag{6.73b}$$

In the above, $k'' = |\boldsymbol{k}''|$, $k' = |\boldsymbol{k}'|$, and c_2 is the light speed in medium 2. Thus, the electric field strength and magnetic flux density in medium 1 are

$$\boldsymbol{E}_1 = \boldsymbol{E} + \boldsymbol{E}'', \quad \boldsymbol{B}_1 = \boldsymbol{B} + \boldsymbol{B}'', \tag{6.74}$$

and those in medium 2 are

$$\boldsymbol{E}_2 = \boldsymbol{E}', \quad \boldsymbol{B}_2 = \boldsymbol{B}'. \tag{6.75}$$

To satisfy the all boundary conditions (6.67)–(6.70) at the interface ($z = 0$) at any time, the phase must be the same for the three waves. This condition is given by

$$\boldsymbol{k} \cdot \boldsymbol{r}\big|_{z=0} = \boldsymbol{k}'' \cdot \boldsymbol{r}\big|_{z=0} = \boldsymbol{k}' \cdot \boldsymbol{r}\big|_{z=0}. \tag{6.76}$$

Equation (6.76) is expressed as

$$\boldsymbol{k} \cdot \boldsymbol{r}_0 = \boldsymbol{k}'' \cdot \boldsymbol{r}_0 = \boldsymbol{k}' \cdot \boldsymbol{r}_0 \tag{6.77}$$

in terms of an arbitrary position vector $\boldsymbol{r}_0$ on the interface. If $\boldsymbol{r}_0$ is given by

$$\boldsymbol{r}_0 = x\boldsymbol{i}_x + y\boldsymbol{i}_y, \tag{6.78}$$

we have

$$k \sin\theta = k'' \sin\theta'' = k' \sin\theta', \tag{6.79}$$

since the wave numbervectors are normal to the y-axis. The speed is the same for the incident and reflected waves in the same medium. Thus, the wavenumbers of these waves are the same ($k = k''$), and we have

$$\theta = \theta''. \tag{6.80}$$

That is, the incident and reflection angles are the same. This is the **law of reflection**. We also have the relationship between the incident and transmission angles as

$$\frac{\sin \theta}{\sin \theta'} = \frac{k'}{k} = \frac{c_1}{c_2} = \left(\frac{\epsilon_2 \mu_2}{\epsilon_1 \mu_1} \right)^{1/2}. \tag{6.81}$$

This is called **Snell's law** for refraction.

In general, the electric field of the incident wave is directed in various directions. Even in such a case, the electromagnetic quantities can be determined using the boundary conditions (6.67)–(6.70).

Exercises

6.1 Solve the problem in Example 6.3 with Eq. (6.3).

6.2 Determine the magnetic energy in a unit length of the concentric supercon-ducting transmission line in Exercise 5.10 in Chap. 5, when we apply current $\pm I$ to it.

6.3 Determine the magnetic field strength in the space between the electrodes of the capacitor in Example 6.5 using the obtained displacement current.

6.4 Show that the equation describing the magnetic flux density in the skin effect is given in the same form of the diffusion equation as Eq. (6.36).

6.5 When we apply an AC electric field parallel to a conductor surface, the internal electric field follows Eq. (6.41). Determine the energy dissipated in a unit surface area in a period of the AC electric field.

6.6 We derived the wave equation for the electric field strength as in Eq. (6.48). Show that the equation of the corresponding magnetic flux density is also the wave equation.

6.7 Determine the value of the electric and magnetic energy densities averaged over one period ($T = 2\pi/\omega$) and over one wave length ($\lambda = 2\pi/k$) for the planar electromagnetic wave given by Eqs. (6.50) and (6.55).

Chapter 7
Direct Current Circuit

Abstract In this chapter, we treat direct current circuits composed of resistors and direct current power sources (voltage source and/or current source), and determine the current and combined resistance. Intuitive and analytic solution methods are described and compared for deeper understanding. The Wheatstone bridge, which is a fundamental bridge circuit to measure unknown resistance, is introduced. We also learn the condition of matching, under which the maximum output power can be obtained from an electric power source with internal resistance.

7.1 Resistance and Ohm's Law

Suppose that when we apply voltage V [V] to a circuit element as shown in Fig. 7.1a, current I [A] flows through it. If V and I obey Ohm's law given by

$$V = RI, \qquad (7.1)$$

this circuit element is called **resistor**. The proportional constant, R, is the **resistance**. When we express this relationship as

$$I = GV, \qquad (7.2)$$

$G = 1/R$ is called the **conductance**. Its unit is [S] (**siemens**).

The resistor is represented by a symbol shown in Fig. 7.1b. The arrow for the voltage V shows that the electric potential at terminal A is measured from the reference point B, and is related to the direction of the current. When the direction of the current is reversed with an unchanged arrow for the voltage, the current is $I = -V/R$.

Suppose that resistors with resistances R_1 and R_2 are connected as shown in Fig. 7.2. This connection is called **connection in series**. It is assumed that current I flows when voltage V is applied between the terminals. In this case the voltage applied to each resistor is $V_1 = R_1 I$ and $V_2 = R_2 I$. So, the total voltage is

© The Author(s), under exclusive license to Springer Nature Switzerland AG 2023   121
T. Matsushita, *Electricity*, https://doi.org/10.1007/978-3-031-44002-1_7

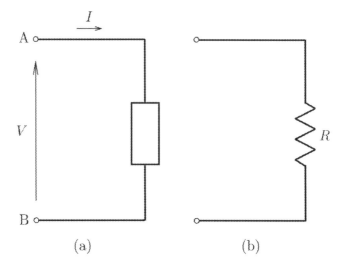

(a)   (b)

Fig. 7.1 a Voltage V applied to a circuit element and current I, and **b** symbol of resistor

Fig. 7.2 Resistors
connected in series

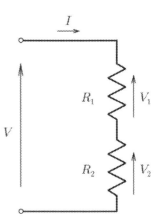

$$V = V_1 + V_2 = (R_1 + R_2)I. \qquad (7.3)$$

Thus, the combined resistance is

$$R = R_1 + R_2. \qquad (7.4)$$

When resistors with resistances $R_1, R_2, \ldots, R_n$ are connected in series, the combined resistance R can be similarly obtained as

Fig. 7.3 Resistors connected in parallel

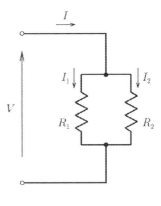

$$R = \sum_{i=1}^{n} R_i. \tag{7.5}$$

The problem of determining the resistance of the slab in Example 3.2 can be regarded as the determination of resistance for resistors connected in series.

Suppose that resistors with resistances R_1 and R_2 are now connected in parallel, as shown in Fig. 7.3. This connection is called **connection in parallel**. It is assumed that current I flows when voltage V is applied between the terminals. In this case, the currents flowing through the resistors are $I_1 = V/R_1$ and $I_2 = V/R_2$. So, the total current is given by

$$I = I_1 + I_2 = \left(\frac{1}{R_1} + \frac{1}{R_2} \right) V. \tag{7.6}$$

Thus, the combined resistance is obtained as

$$\frac{1}{R} = \frac{1}{R_1} + \frac{1}{R_2}. \tag{7.7}$$

When resistors with resistances $R_1, R_2, \ldots, R_n$ are connected in parallel, the combined resistance R can be similarly obtained as

$$\frac{1}{R} = \sum_{i=1}^{n} \frac{1}{R_n}. \tag{7.8}$$

The problem of determining the resistance of the quarter circular prism in Example 3.3 can be regarded as determination of the combined resistance for resistors connected in parallel.

Example 7.1 Resistors are connected as shown in Fig. 7.4. Determine the combined resistance between the two terminals.

Fig. 7.4 Combined resistor
composed of three resistors

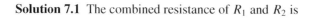

Solution 7.1 The combined resistance of R_1 and R_2 is

$$R_{12} = \left(\frac{1}{R_1} + \frac{1}{R_2} \right)^{-1} = \frac{R_1 R_2}{R_1 + R_2}.$$

The total resistance is the combined resistance of R_{12} and R_3 connected in series, and we have

$$R = R_{12} + R_3 = \frac{R_1 R_2 + R_2 R_3 + R_3 R_1}{R_1 + R_2}.$$

$\diamondsuit$

Example 7.2 The total combined resistance in Fig. 7.5 is 6 Ω. Determine the resistance R.

Fig. 7.5 Combined resistor
composed of 5 resistors

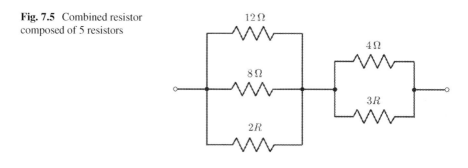

Solution 7.2 The combined resistance of the three resistors on the left side is

$$\frac{1}{R_1} = \frac{1}{12} + \frac{1}{8} + \frac{1}{2R} = \frac{5R + 12}{24R}.$$

This yields $R_1 = 24R/(5R + 12)$. The combined resistance of the two resistors on the right side is $R_2 = 12R/(3R + 4)$. So, the given condition is

$$\frac{24R}{5R + 12} + \frac{12R}{3R + 4} = 6.$$

This leads to $7R^2 - 16R - 48 = 0$, or

$$(7R + 12)(R - 4) = 0.$$

From the condition $R > 0$, we have $R = 4[\Omega]$.

$\diamond$

7.2 Electric Power Source

An **electric power source** is an element of electric circuits that has electromotive force. There are direct voltage sources and direct current sources in direct current (DC) circuits. The electric power source in alternating current (AC) circuits will be mentioned in Chap. 8.

7.2.1 Direct Voltage Source

When a load is connected to the two terminals, a **direct current (DC) voltage source** provides a constant electromotive force with time independently of the resistance of the load. The symbol of a DC voltage source is shown in Fig. 7.6a, where E is the electromotive force and the figure shows that the electric potential is higher at terminal A. Although it may be consistent to use V_{em} for the electromotive force as done in Chap. 3, we use E from now on, based on the custom for electric circuits. Note, however, that it is different from the electric field strength. When a resistor with resistance R is connected to a direct voltage source of electromotive force E, as shown in Fig. 7.6b, the current that flows through the resistor is

$$I = \frac{E}{R}. \tag{7.9}$$

In this case, the voltage between the two terminals of the resistor, i.e., the voltage drop is

$$V = RI = E, \tag{7.10}$$

Fig. 7.6 a Symbol of DC
voltage source and **b** the case
in which a DC voltage source
is connected to a resistor

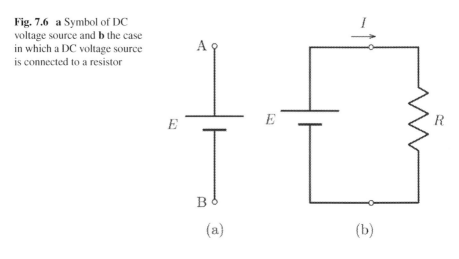

as shown in Fig. 7.7, and it is equal to the electromotive force of the DC voltage
source.

In practical cases, when the resistance of the load becomes small, the relation
given by Eq. (7.9) does not hold, and the current becomes smaller than the expected
value as

$$I = \frac{E}{R + R_0}. \tag{7.11}$$

This is because there is a power loss even in the electric power source through which
current flows, and R_0 in Eq. (7.11) is called the **internal resistance**. The equivalent
circuit of a DC voltage source with an internal resistance is shown in Fig. 7.8. An
ideal DC voltage source without internal resistance cannot be shunted, as shown
in Fig. 7.9a, since the current diverges. When the voltage source contains internal
resistance, it is possible, as shown in Fig. 7.9b. When a resistor with resistance R is
connected to a DC voltage source of electromotive force E and internal resistance
R_0, as shown in Fig. 7.10, the current is given by Eq. (7.11), and the voltage across
the resistor is

Fig. 7.7 Voltage on a
resistor and direction of the
current

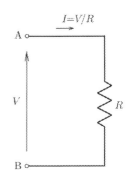

$$V = \frac{E}{1 + (R_0/R)}.$$ (7.12)

Thus, the internal resistance must be sufficiently smaller than the load resistance to realize an ideal voltage source that can apply voltage E to the load resistor. Or, the resistance of the load resistor must be sufficiently larger than the internal resistance of the voltage source to use it as an ideal source.

We do not suppose that there is internal resistance in DC voltage sources, except in the designated cases.

Fig. 7.8 Equivalent circuit of DC voltage source with internal resistance

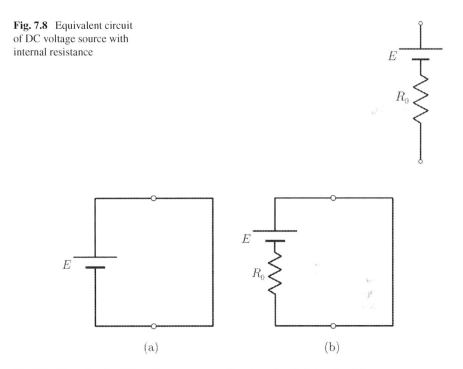

(a)   (b)

Fig. 7.9 Short-circuit of DC voltage source **a** without and **b** with internal resistance

Fig. 7.10 Resistor connected to DC voltage source with internal resistance

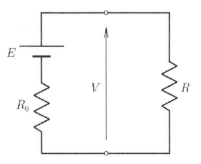

7.2.2   *Direct Current Source*

When a load is connected to the two terminals, a **direct current (DC) source** provides a constant current with time independently of the resistance of the load. The symbol of a DC current source is shown in Fig. 7.11a, where J is the current supplied by the current source, and the figure shows that the current flows from B to A. When a resistor of resistance R is connected to a DC current source, as shown in Fig. 7.11b, the voltage across the resistor is

$$V = RJ. \tag{7.13}$$

When the load resistance becomes large, it is not possible to give a constant current J. In an extreme case where the two terminals of the current source are opened, current must be zero. So, we assume also an internal resistance R_0 for direct current source, as shown in Fig. 7.12. Suppose that a resistor of resistance R is connected to the DC current source, as shown in Fig. 7.13. We denote the current flowing through the resistor by I, while the current flowing through the internal resistance is $J - I$. Since the voltage across each resistance is the same, we have $RI = R_0(J - I)$. Then, the current is given by

$$I = \frac{R_0 J}{R + R_0} = \frac{J}{1 + (R/R_0)}. \tag{7.14}$$

Thus, the internal resistance must be sufficiently larger than the load resistance to realize an ideal current source that can apply current J to the load resistor, or the resistance of load resistor must be sufficiently smaller than the internal resistance of the current source to use it as an ideal source.

Fig. 7.11 a Symbol of DC current source and **b** the case in which a DC current source is connected to a resistor

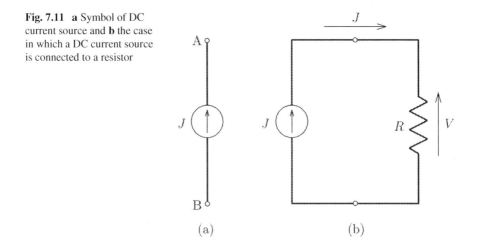

(a)   (b)

Fig. 7.12 Equivalent circuit of DC current source with internal resistance

Fig. 7.13 Resistor connected to DC current source with internal resistance

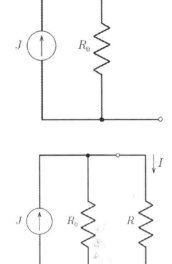

Comparing Eq. (7.14) with Eq. (7.11) for the DC voltage source, we find that two sources are equivalent to each other when R_0 is common and the following relation holds:

$$E = R_0 J. \tag{7.15}$$

In fact, the current given by Eq. (7.11) becomes $I = E/R_0 = J$ for a DC voltage source with internal resistance sufficiently larger than the load resistance, and this voltage source acts as a current source.

We do not suppose internal resistance in DC current sources either, except in the designated cases.

7.3 Resistor Circuit

Here, we treat the cases of circuits composed of resistors and DC power sources and try to determine currents or voltages. Suppose the circuit shown in Fig. 7.14. Since voltage E is applied between A and D, the current flowing through point B is

$$I_1 = \frac{E}{R_1 + R_2}, \tag{7.16}$$

Fig. 7.14 Resistor circuit
composed of four resistors

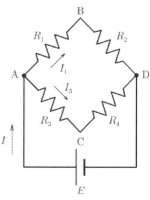

and the voltages across the resistors of resistances R_1 and R_2 are respectively given
by

$$V_1 = R_1 I_1 = \frac{R_1 E}{R_1 + R_2}, \quad V_2 = R_2 I_1 = \frac{R_2 E}{R_1 + R_2}. \tag{7.17}$$

The current flowing through point C is

$$I_3 = \frac{E}{R_3 + R_4}, \tag{7.18}$$

and the voltages across R_3 and R_4 are respectively given by

$$V_3 = R_3 I_3 = \frac{R_3 E}{R_3 + R_4}, \quad V_4 = R_4 I_3 = \frac{R_4 E}{R_3 + R_4}. \tag{7.19}$$

We shall determine the voltage between B and C, i.e., the electric potential at point
B measured from point C. The electric potential at point D measured from point C
is $-V_4$, and the electric potential at point B measured from point D is V_2. So, the
voltage between B and C is determined to be

$$\begin{aligned}
V_{\mathrm{BC}} = V_2 - V_4 &= \frac{[R_2(R_3 + R_4) - R_4(R_1 + R_2)]E}{(R_1 + R_2)(R_3 + R_4)} \\
&= \frac{(R_2 R_3 - R_4 R_1)E}{(R_1 + R_2)(R_3 + R_4)}.
\end{aligned} \tag{7.20}$$

The circuit shown in Fig. 7.14 is the **Wheatstone bridge**. The condition that the
voltage between B and C becomes zero when the following condition is satisfied:

$$R_2 R_3 = R_4 R_1. \tag{7.21}$$

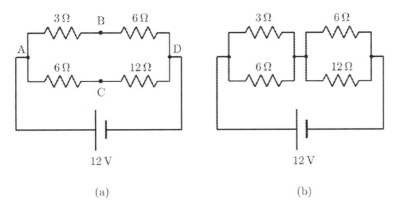

Fig. 7.15 Resistance circuit composed of four resistors. **b** is the case in which points B and C are short-circuited in **a**

Suppose that resistances R_1 and R_2 are known and R_4 is unknown. If we adjust R_3 so that current does not flow through BC, we can determine R_4 from Eq. (7.21) by substituting the values of R_1, R_2, and R_3. The Wheatstone bridge is used for the measurement of unknown resistances in this way.

Here, we compare the two resistor circuits shown in Fig. 7.15a, b. Then, it can be easily found that the voltages across and the current flowing through each resistor are the same. The circuit in Fig. 7.15a has the same structure as that in Fig. 7.14, and the circuit in Fig. 7.15b is realized by connecting points B and C in the circuit in Fig. 7.15a. It can be seen that the voltage between B and C is zero, using Eq. (7.21). This is the reason why current does not flow between B and C, even if the two points are connected. In addition, the same result is obtained even if B and C are connected through a resistor with any resistance.

Example 7.3 Determine the current flowing through the resistor with resistance R when current I is 1.5 A in the circuit shown in Fig. 7.16.

Fig. 7.16 Resistor circuit with unknown resistance

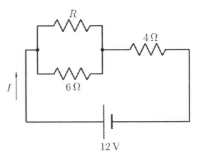

Solution 7.3 Since the voltage across the 4 Ω resistor is 6 V, the other 6 V is applied across the 6 Ω resistor. Hence, current of 1 A flows through this resistor. This means that the current that flows through resistor R is 0.5 A. Thus, we can obtain the solution without determining the unknown resistance R, which is 12 Ω. It is possible to solve the problem after determining R. In this sense, this problem is imperfect. On the other hand, we can effectively solve the problem, as shown here. It is important to see the essential point without being affected by custom.

◇

Example 7.4 When the voltage across the $2R$ resistor is 8 V in the circuit shown in Fig. 7.17, determine the value of resistance R.

Fig. 7.17 Resistor circuit including unknown resistances

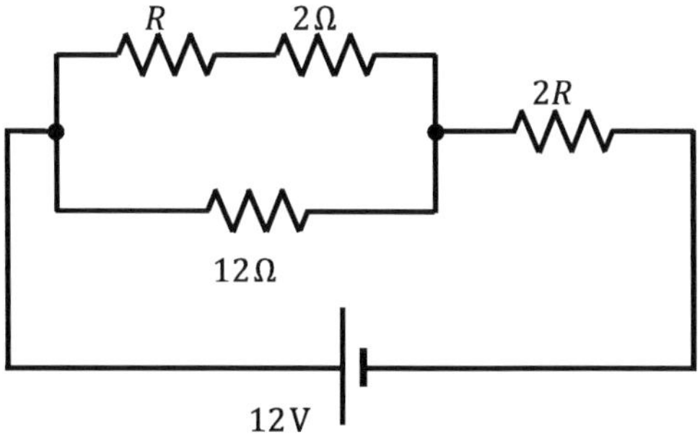

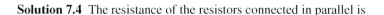

Solution 7.4 The resistance of the resistors connected in parallel is

$$R' = \frac{12(R+2)}{R+14},$$

and the voltage of 4 V is applied to it. Then, the voltage across $2R$ is 8 V, and R' must be equal to R. Thus, we obtain $R^2+2R-24 = 0$. This leads to $(R+6)(R-4) = 0$, and we have $R = 4[\Omega]$ from the condition $R > 0$.

◇

Here, we shall determine the combined resistance between A and B in the tetra-hedral resistor circuit composed of 6 resistors with resistance R shown in Fig. 7.18. If we find that the voltage between C and D is zero due to symmetry when we apply a voltage between A and B, we can solve this problem easily. Since current does not flow through the resistor between C and D, we can remove it. Then, three resistor branches with resistances R, $2R$ through C, and $2R$ through D are connected in parallel between A and B. Thus, the combined resistance is determined to be $R/2$.

Fig. 7.18 Resistors arranged
in each edge of a tetrahedron

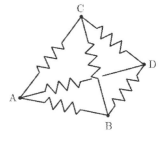

Example 7.5 Determine the combined resistance between A and B for the twelve
resistors with resistance R shown in Fig. 7.19.

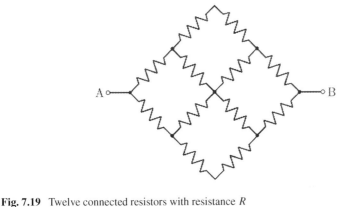

Fig. 7.19 Twelve connected resistors with resistance R

Solution 7.5 It can be shown from symmetry that the resistance does not change,
even when the connection is changed, as shown in Fig. 7.20. The connection in
which points C and D are short-circuited is the given connection. This is because
the electric potential difference is zero between C and D. Since the resistance of the

Fig. 7.20 Connected
resistors equivalent to those
shown in Fig. 7.19. When C
and D are short-circuited, the
connected resistors shown in
Fig. 7.19 is obtained

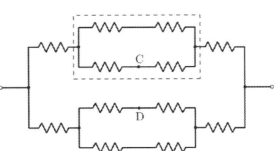

part surrounded by the dotted line is equal to R, the combined resistance is two $3R$ resistors connected in parallel. Thus, the combined resistance is determined to be $3R/2$.

◇

We solved the problem in Example 7.5 using a symmetry condition. Such a method is not always successful, however. Then, it is necessary to solve the problem analytically with unknown variables. Suppose that voltage V is applied between the two terminals. Then, three currents are used as unknown variables, as shown in Fig. 7.21.

There is a relation, $I_1 = I_2 + I_3$, among the three currents I_1, I_2, and I_3. The current flowing through branch AC must be I_2, and that flowing through branch BC must be I_3, since no current flows from the lower half from symmetry. Since the voltage between C$'$ and C through A and that through B is common, we see that $I_2 = I_3$. Thus, the voltage between D$'$ and D through A is given by

$$V = RI_1 + 2RI_2 + RI_1 = 3RI_1. \tag{7.22}$$

Since the total current in this resistor circuit is $I = 2I_1$, the resistance of the circuit is determined to be

$$\frac{V}{I} = \frac{3}{2}R. \tag{7.23}$$

It may be necessary to use many unknown variables depending on the problems. We discuss the number of necessary variables in Chap. 11.

Next, we shall determine the current flowing through branch BC in the circuit shown in Fig. 7.22. In this case the voltage between B and C is unknown, and it is not easy to determine the current. So, we use unknown variables. We denote by I_1 and I_3 the currents flowing through the resistors R_1 and R_3, respectively. The current I is defined to flow from B to C. Then, from the continuity of currents, the currents I_2 and I_4 flowing through the resistors of R_2 and R_4 are respectively given by

$$I_2 = I_1 - I, \quad I_4 = I_3 + I. \tag{7.24}$$

Fig. 7.21 Currents flowing through resistors

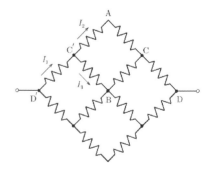

Fig. 7.22 Resistor circuit

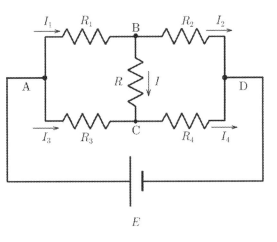

Since all the currents flowing through each resistor can be given, the voltages across each resistor can be obtained. Since the unknown variables are I, I_1, and I_3, we need three independent conditions. First, the voltage between A and D through B is equal to E. Thus, we have

$$R_1 I_1 + R_2(I_1 - I) = E. \tag{7.25}$$

Next, the voltage between A and D through C is also equal to E:

$$R_3 I_3 + R_4(I_3 + I) = E. \tag{7.26}$$

Another condition is for the voltage between A and C, which is given by

$$R_1 I_1 + RI = R_3 I_3. \tag{7.27}$$

We have other conditions. The number of independent conditions is three, however, and we can solve the problem with the above conditions.

Using Eq. (7.25), I_1 is expressed as

$$I_1 = \frac{E + R_2 I}{R_1 + R_2}, \tag{7.28}$$

and using Eq. (7.26), I_3 is expressed as

$$I_3 = \frac{E - R_4 I}{R_3 + R_4}. \tag{7.29}$$

Substitution of these equations into Eq. (7.27) leads to

$$I = \frac{(R_2 R_3 - R_4 R_1)E}{R_1 R_2 (R_3 + R_4) + R_3 R_4 (R_1 + R_2) + R(R_1 + R_2)(R_1 + R_2)}. \quad (7.30)$$

Thus, if the condition $R_2 R_3 = R_4 R_1$ is satisfied, the voltage between B and C is zero, and current does not flow through the resistor R in this case also.

Problems can be solved effectively using unknown variables, as shown above. This method is useful for complicated circuits. Such solution methods are systematically introduced in Chap. 11.

Example 7.6 Determine the resistance between A and B in the connected resistors shown in Fig. 7.23. (*Hint*: If we denote this resistance by R, the resistance on the right side of A′B′ is also R.)

Fig. 7.23 Connected resistors

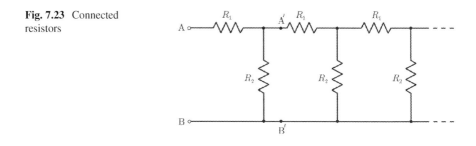

Solution 7.6 We denote by R the combined resistance. Then, this is equal to the combined resistance of the connected resistors shown in Fig. 7.24. Thus, we have

$$R = R_1 + \frac{R_2 R}{R + R_2}.$$

This leads to $R^2 - R_1 R - R_1 R_2 = 0$, and the resistance is determined to be

Fig. 7.24 Equivalent connected resistors

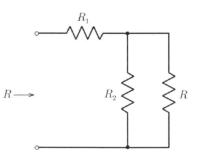

$$R = \frac{1}{2}\left[R_1 + \left(R_1^2 + 4R_1 R_2\right)^{1/2}\right].$$

◇

7.4   Electric Power in Resistor

In the condition where current I flows through a resistor of resistance R under voltage V, the electric power given by Eq. (3.22) is not stored as energy but is dissipated, resulting in a Joule heat. That is, the electric power in this case is the energy dissipated in unit time (1 s).

Suppose that a resistor R is connected to a DC voltage source with electromotive force E and internal resistance R_0, as shown in Fig. 7.25. The current flowing through the resistor is $I = E/(R + R_0)$ and the electric power is

$$P = RI^2 = \frac{RE^2}{(R + R_0)^2}. \tag{7.31}$$

The condition in which the dissipated power is at a maximum under variation in R is obtained from

$$\frac{\mathrm{d}P}{\mathrm{d}R} = 0 \tag{7.32}$$

to be

$$R = R_0. \tag{7.33}$$

In this case, we can see that P takes on its maximum value from $\mathrm{d}^2 P/\mathrm{d}R^2 = -E^2/(8R_0^3) < 0$. Equation (7.33) gives the condition in which the maximum power is obtained, and this condition is called **matching**. A part of the electric power is dissipated in the power source due to the internal resistance R_0.

Fig. 7.25 Resistor
connected to DC voltage
source with internal
resistance

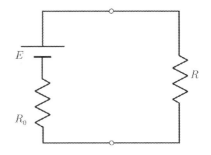

Example 7.7 Determine the value of R that provides the maximum electric power dissipated in resistor R in the circuit shown in Fig. 7.26.

Fig. 7.26 Resistor circuit

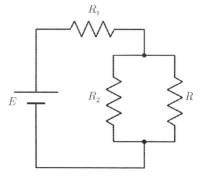

Solution 7.7 The combined resistance is R_1 and $R_2 R/(R + R_2)$ connected in series. So, the voltage applied to R_2 is

$$V = E \cdot \frac{R_2 R/(R + R_2)}{R_1 + R_2 R/(R + R_2)} = \frac{R_2 R E}{R_1 R_2 + (R_1 + R_2) R}.$$

The dissipated power in R is given by

$$P = \frac{V^2}{R} = \frac{R_2^2 R E^2}{[R_1 R_2 + (R_1 + R_2) R]^2}.$$

Then, we have

$$\frac{\mathrm{d}P}{\mathrm{d}R} = \frac{R_2^2 E^2 [R_1 R_2 - (R_1 + R_2) R]}{[R_1 R_2 + (R_1 + R_2) R]^3}.$$

The condition $\mathrm{d}P/\mathrm{d}R = 0$ yields

$$R = \frac{R_1 R_2}{R_1 + R_2}.$$

In this case, we can show that $\mathrm{d}^2 P/\mathrm{d}R^2 < 0$, and the above condition is the solution.

Exercises

7.1 Determine the combined resistance of the connected resistors shown in Fig. 7.27.

Fig. 7.27 Connected resistors

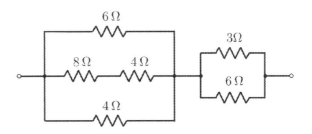

7.2 When we exchange R and $R' = 12\ \Omega$ in Fig. 7.28, the combined resistance changes to a half value before the exchange. Determine R.

Fig. 7.28 Connected resistors with unknown resistance R

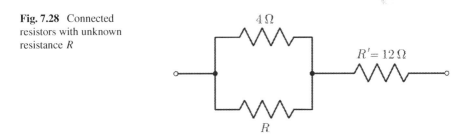

7.3 The total current I flowing in the circuit shown in Fig. 7.29 is 1 A. Determine the resistance R.

Fig. 7.29 Resistor circuit with unknown resistances

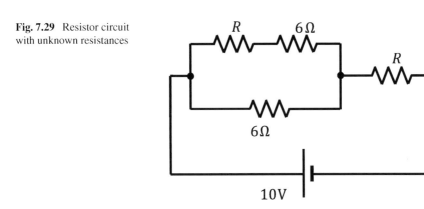

Fig. 7.30 Currents flowing
through connected resistors

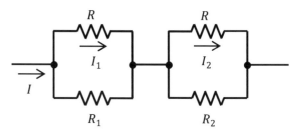

7.4 Determine the resistance R in the connected resistors shown in Fig. 7.30,
     when the condition $I_1 = 2I_2$ is satisfied. Derive the condition for R_1 and R_2
     for which such a solution exists.

7.5 Determine the combined resistance of the connected resistors shown in
     Fig. 7.31. The resistance of each resistor is R.

7.6 Determine the combined resistance shown in Fig. 7.31 using unknown
     variables.

Fig. 7.31 Connected
resistors

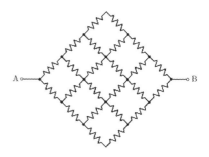

Fig. 7.32 Connected
resistors

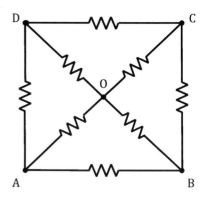

Fig. 7.33 Connected
resistors

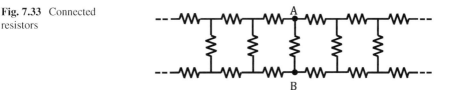

7.7 Determine the combined resistances between A and B, A and C, and O and A
for the connected resistors shown in Fig. 7.32. The resistance of each resistor
is R.

7.8 Determine the combined resistance between A and B for the connected
resistors shown in Fig. 7.33. The resistance of each resistor is R.

7.9 Determine the value of R for which the power dissipated in the resistor takes
on its maximum value in the circuit shown in Fig. 7.34.

Fig. 7.34 Resistor circuit
with DC current source

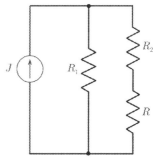

7.10 Determine the ratio of the resistances, R_1/R_2, when current I_2 is at a minimum
under the condition $R_1 + R_2 = C$ (const.) in the circuit shown in Fig. 7.35.

Fig. 7.35 Resistor circuit
with DC voltage source

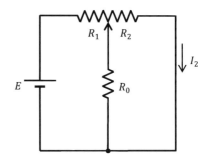

7.11   Figure 7.36 is a resistor circuit in which resistance R_1 can be varied. Derive
the condition for R_1 for which the electric power in R_1 takes on its maximum
value and determine the maximum electric power.

Fig. 7.36 Resistor circuit
with DC voltage source

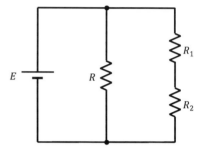

Chapter 8
Transient and Steady Responses of Electric Circuits

Abstract In this chapter we learn the relationships between the applied voltage and the current for each element of the circuit, i.e., resistor, capacitor, and inductor coil. Then, we learn analytic methods to solve the response of the circuit composed of these elements, such as the transient response when a DC voltage source is switched on and the steady response when an AC voltage source is connected to the circuit. We also treat the energy in each circuit element. The energy stored or dissipated in each circuit element in the transient and steady states will be determined.

8.1 Each Circuit Element

The resistor, DC voltage source, and DC current source were treated in the last chapter. There are other elements that compose electric circuits: These include capacitor to store electric charge, which we learned about in Chap. 2 and coil to store magnetic flux, which were discussed in Chap. 5. AC voltage sources and AC current sources are also included. The fundamental properties of these elements are discussed in this chapter. Since a linear relationship holds between the voltage and current for each element, these are called **linear circuit elements**. This does not necessarily mean a proportional relationship, however.

8.1.1 Resistor

The **resistor** is a circuit element that has **resistance** R, as introduced in Sect. 7.1. When a voltage $V(t)$ is applied to it, as shown in Fig. 8.1, the relationship holds between the voltage and current $I(t)$ as

$$V(t) = RI(t) \tag{8.1}$$

T. Matsushita, *Electricity*,
https://doi.org/10.1007/978-3-031-44002-1_8

Fig. 8.1 Resistor

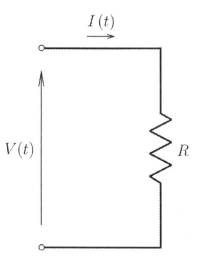

or

$$I(t) = R^{-1}V(t) = GV(t), \tag{8.2}$$

including the case in which these change with time.

8.1.2   Capacitor

The **capacitor** is a circuit element to store electric charge, as mentioned in Sect. 2.2. When the electric charges on the two electrodes are $\pm Q(t)$, the voltage between the electrodes is

$$V(t) = \frac{Q(t)}{C}, \tag{8.3}$$

where C is the **capacitance**. As mentioned in Sect. 3.1, the change in the electric charge $Q(t)$ is related to the current $I(t)$ as

$$I(t) = \frac{dQ(t)}{dt}. \tag{8.4}$$

Thus, the relationship between the voltage $V(t)$ and the current $I(t)$ in Fig. 8.2 is given by

Fig. 8.2 Capacitor

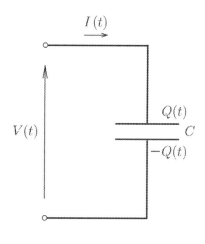

$$V(t) = \frac{1}{C} \int_{t_0}^{t} I(t)dt \qquad (8.5)$$

or

$$I(t) = C\frac{dV(t)}{dt}. \qquad (8.6)$$

In Eq. (8.5), t_0 is the time at which $V(t_0) = 0$ is satisfied.

8.1.3 Coil

The **coil** or **inductor** is a circuit element used to store magnetic flux, as treated in Sect. 5.2. When the current is $I(t)$, the magnetic flux that penetrates the coil of **inductance** L is given by

$$\Phi(t) = LI(t). \qquad (8.7)$$

For the coil shown in Fig. 8.3, the voltage that we must apply to maintain the current against the electromotive force induced by the varying current is

$$V(t) = \frac{d\Phi(t)}{dt} = L\frac{dI(t)}{dt}, \qquad (8.8)$$

or the current is given by

Fig. 8.3 Coil

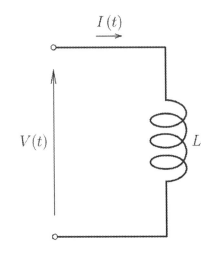

$$I(t) = \frac{1}{L} \int_{t_0}^{t} V(t)dt, \tag{8.9}$$

where t_0 is the time at which $I(t_0) = 0$ is satisfied.

8.1.4 Voltage Source

The **voltage source** is a circuit element that supplies a voltage equal to the electro-motive force $E(t)$ independently of the circuit element connected to it, as shown in Fig. 8.4, and the relation holds:

$$V(t) = E(t). \tag{8.10}$$

Fig. 8.4 Voltage source

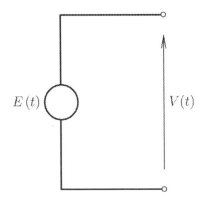

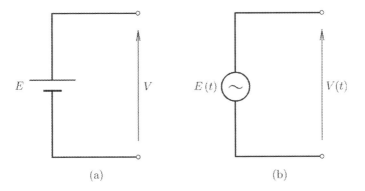

(a)   (b)

Fig. 8.5 Commonly used voltage sources: **a** DC voltage source and **b** AC voltage source

The current flowing through depends on the circuit element that is connected to the source. There are DC voltage sources that provide a constant direct voltage, as shown in Fig. 8.5a and AC voltage sources that provide a sinusoidal voltage, as shown in Fig. 8.5b.

8.1.5 Current Source

The **current source** is a circuit element that supplies a fixed value of the current $J(t)$, independently of the circuit element connected to it, as shown in Fig. 8.6, and the relation holds:

$$I(t) = J(t). \tag{8.11}$$

Fig. 8.6 Current source

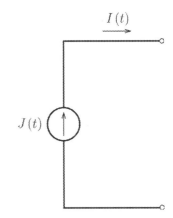

The voltage across the connected circuit element depends on the element itself. There are DC current sources that provide a constant direct current and AC current sources that provide a sinusoidal current.

8.2   Electric Power and Energy in Circuit Elements

We will derive the electric power and energy in each circuit element, except for power sources.

8.2.1   Resistor

Suppose that we apply voltage $V(t)$ across a resistor with resistance R. The current flowing through the resistor is $I(t) = V(t)/R$, and the electric power, i.e., the electric energy, that goes into the resistor in unit time is given by

$$P(t) = V(t)I(t) = RI^2(t) = \frac{V^2(t)}{R}. \tag{8.12}$$

For example, in the case of DC voltage, a constant electric energy continues to penetrate the resistor, but the electric state does not change. So, this energy is dissipated. In fact, the energy changes to a Joule heat and the law of conservation of energy holds in a wide sense.

Example 8.1 Determine the dissipated energy in one period $T = 2\pi/\omega$ when AC voltage $V(t) = V_0 \cos \omega t$ is applied to a resistor with resistance R.

Solution 8.1 The dissipated electric power is obtained from Eq. (8.12) as $P(t) = \left(V_0^2/R\right) \cos^2 \omega t$ and the dissipated energy in one period is determined to be

$$W = \int_0^T \frac{V_0^2}{R} \cos^2 \omega t \, dt = \frac{V_0^2}{2R} T = \frac{1}{2} R I_0^2 T,$$

where $I_0 = V_0/R$ is the amplitude of the current. This value is the same as when DC voltage $V_e = V_0/\sqrt{2}$ or DC current $I_e = I_0/\sqrt{2}$ is applied to the resistor. The values of the voltage and current, V_e and I_e, are called **effective values**.

$\diamondsuit$

8.2.2 *Capacitor*

Suppose that we apply voltage $V(t)$ across a capacitor with capacitance C. The electric power that goes into the capacitor is given by

$$P(t) = V(t)I(t) = CV(t)\frac{dV(t)}{dt}. \tag{8.13}$$

If the initial condition is $V(t_0) = 0$, the electric energy that goes into the capacitor until the voltage reaches $V(t)$ at time t is

$$W(t) = C\int_{t_0}^{t} V(t)\frac{dV(t)}{dt}dt = C\int_{0}^{V(t)} VdV = \frac{1}{2}CV^2(t), \tag{8.14}$$

which agrees with the electrostatic energy given by Eq. (2.42). If the variation stops when the voltage reaches V, the energy flow into the capacitor stops, and the electric energy is stored in the capacitor. When the voltage across the capacitor is reduced to zero, the electric power is negative, the energy comes back to the power source, and the electric energy in the capacitor decreases to zero. Thus, the energy can be stored in the capacitor but is not dissipated. This is essentially different from the case of the resistor.

8.2.3 *Coil*

Suppose that we apply current $I(t)$ to a coil with inductance L. The electric power that goes into the coil is given by

$$P(t) = V(t)I(t) = LI(t)\frac{dI(t)}{dt}. \tag{8.15}$$

If the initial condition is $I(t_0) = 0$, the electric energy that goes into the coil until the current reaches $I(t)$ at time t is

$$W(t) = L\int_{t_0}^{t} I(t)\frac{dI(t)}{dt}dt = I\int_{0}^{I(t)} IdI = \frac{1}{2}LI^2(t), \tag{8.16}$$

which agrees with the magnetic energy given by Eq. (6.15). If the variation stops when the current reaches I, the energy flow into the coil stops, and the electric energy is stored in the coil. When the current is reduced to zero, the electric power is negative, the energy comes back to the power source, and the electric energy in the coil decreases to zero. Thus, the energy can be stored in the coil but is not dissipated, similarly to capacitors.

8.3 *LRC* Circuit

Suppose a typical electric circuit in which a coil, resistor, and capacitor are connected in series, as shown in Fig. 8.7. The voltage drop due to each circuit element is given by Eqs. (8.8), (8.1), and (8.5), and the balance with the electromotive force of the voltage source is given by

$$E(t) = L\frac{dI(t)}{dt} + RI(t) + \frac{1}{C}\int I(t)dt. \tag{8.17}$$

The form of this equation is well known as a dynamic equation of motion for mechanical systems. Suppose the case in which a solid body of mass m connected to a spring with spring constant k is moving in a viscous medium of viscosity η, as shown in Fig. 8.8. The equation of motion of this matter is given by

$$m\frac{d^2x}{dt^2} + \eta\frac{dx}{dt} + kx = f, \tag{8.18}$$

where f is an external force and x is the displacement of the mass from its equilibrium position. The first term is the force needed for acceleration, the second term is the viscous force, and the third term is the spring force. If we use the velocity of the mass,

Fig. 8.7 Electric circuit composed of coil, resistor, capacitor, and voltage source

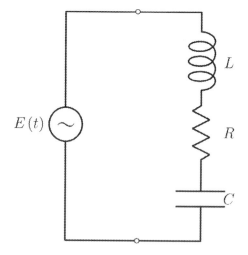

Fig. 8.8 Mechanical
system, composed of spring,
mass, and viscous medium

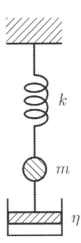

$$v(t) = \frac{dx}{dt},\tag{8.19}$$

instead of the displacement x, Eq. (8.18) leads to

$$f = m\frac{dv(x)}{dt} + \eta v(t) + k\int v(t)dt.\tag{8.20}$$

So, this equation has the same form as Eq. (8.17). Thus, it is effective for under-standing various phenomena to know the property of this electric circuit. Here, we will investigate the behavior, especially the transient response and steady response, of the electric circuit.

8.3.1 Transient Response

We treat the variation of the current shown in Fig. 8.9 after turning switch S on to apply the voltage to the circuit. The equation of the circuit for $t \geq 0$ is

$$L\frac{dI(t)}{dt} + RI(t) + \frac{1}{C}\int_0^t I(t)dt = E.\tag{8.21}$$

Here, we assume a general solution of the equation in which the right side is assumed to be zero as

$$I(t) = Ke^{St}.\tag{8.22}$$

Fig. 8.9 Electric circuit composed of coil, resistor, capacitor, and DC voltage source. Switch S is turned on at $t = 0$

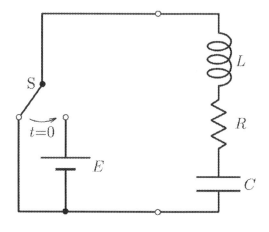

Then, the time differentiation and time integration become multiplication by S and S^{-1}, respectively. Hence, this equation leads to

$$\left(LS + R + \frac{1}{CS}\right)K e^{St} = 0. \tag{8.23}$$

Thus, we have

$$LS + R + \frac{1}{CS} = 0, \tag{8.24}$$

and S is determined to be

$$S = \frac{-R \pm \left[R^2 - 4(L/C)\right]^{1/2}}{2L} \equiv S_{1,2}. \tag{8.25}$$

The solution is given in the form:

$$I(t) = K_1 e^{S_1 t} + K_2 e^{S_2 t}. \tag{8.26}$$

The initial condition to be satisfied is

$$I(0) = K_1 + K_2 = 0. \tag{8.27}$$

Another condition to be satisfied is

$$\int_0^\infty I(t)dt = CE \tag{8.28}$$

after a sufficiently long time at which $I(t)$ reduces to zero. This yields:

$$\frac{K_1}{S_1} + \frac{K_2}{S_2} = -CE. \tag{8.29}$$

Using the above two conditions, the coefficients are determined to be

$$K_1 = -K_2 = \frac{E}{\left[R^2 - 4(L/C)\right]^{1/2}}. \tag{8.30}$$

Here, we assume that L and C are constant and only R changes. For $R < 2(L/C)^{1/2}$, the term inside the square root is negative, and hence, S_1 and S_2 are complex numbers, and K_1 and K_2 are imaginary numbers. If we denote as $\left[R^2 - 4(L/C)\right]^{1/2}/(2L) = i\alpha$, the current leads to

$$I(t) = \frac{E}{2i\alpha L} \exp\left(-\frac{Rt}{2L}\right)\left(e^{i\alpha t} - e^{-i\alpha t}\right) = \frac{E}{\alpha L} \exp\left(-\frac{Rt}{2L}\right) \sin \alpha t. \tag{8.31}$$

The variation of the current with time is shown in Fig. 8.10. The current oscillates with angular frequency α and its amplitude decreases in proportion to $\exp[-(R/2L)t]$. Such an oscillation is called **damped oscillation**.

When $R > 2(L/C)^{1/2}$, S_1 and S_2 are negative real numbers, and K_1 and K_2 are real numbers. If we define $\left[R^2 - 4(L/C)\right]^{1/2}/(2L) = \beta$, the current is given by

$$I(t) = \frac{E}{2\beta L}\left\{\exp\left[-\left(\frac{R}{2L} - \beta\right)t\right] - \exp\left[-\left(\frac{R}{2L} + \beta\right)t\right]\right\}. \tag{8.32}$$

The variation of the current with time in this case is shown in Fig. 8.11. The current increases initially, and then, it decreases exponentially without oscillation. Formally

Fig. 8.10 Damped oscillation of current

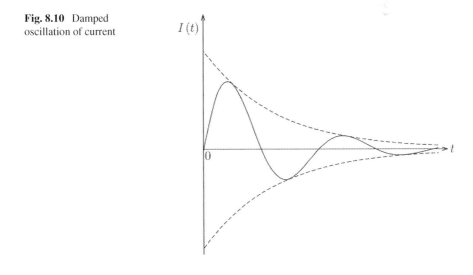

Fig. 8.11 Variation of the current when K_1 and K_2 are real numbers

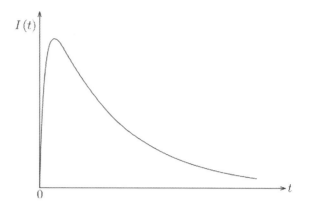

this solution corresponds to that given by Eq. (8.31) with replacements of α and $\sin \alpha t$ by β and $\sinh \beta t$, respectively.

In the case of $R = 2(L/C)^{1/2}$, the denominator of K_1 and K_2 is zero, and the above general solution cannot be used. So, we must solve Eq. (8.21) again. Among the general solution of the form of $(Kt + K')e^{St}$, we choose

$$I(t) = K t e^{St}, \tag{8.33}$$

which satisfies the initial condition, $I(0) = 0$. Substitution of this into Eq. (8.21) leads to

$$K\left\{ \left[S + \frac{2}{(L/C)^{1/2}} + \frac{1}{SLC} \right] t + 1 - \frac{1}{LCS^2} \right\} e^{St} + \frac{K}{LCS^2} = \frac{E}{L}. \tag{8.34}$$

From this condition, we have

Fig. 8.12 Variation of the current at critical damping

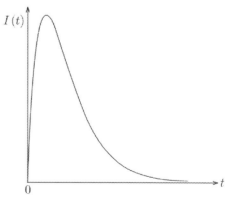

$$S = -\frac{1}{(L/C)^{1/2}}, \quad K = \frac{E}{L}, \tag{8.35}$$

and the solution is obtained:

$$I(t) = \frac{E}{L} t \exp\left[-\frac{t}{(L/C)^{1/2}}\right]. \tag{8.36}$$

The variation of the current with time is shown in Fig. 8.12. This condition of R is called **critical damping**.

Example 8.2 Suppose that DC voltage E is applied by turning switch S on at $t = 0$ in the electric circuit with a resistor and capacitor shown in Fig. 8.13. Determine the variation of the current with time.

Fig. 8.13 Electric circuit composed of resistor, capacitor, and DC voltage source. DC voltage is applied by turning switch S on at $t = 0$

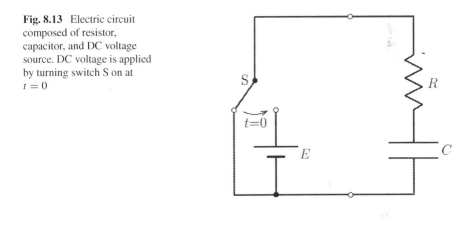

Solution 8.2 The equation for the given circuit is

$$RI(t) + \frac{1}{C}\int_0^t I(t)dt = E.$$

If we put $I(t) = I_0 e^{St}$, the above equation leads to

$$RI_0 e^{St} + \frac{I_0}{CS}(e^{St} - 1) = E.$$

Thus, we have $S = -1/(RC)$ and $I_0 = E/R$ from the coefficient of the term e^{St} and the constant term, respectively. Thus, the current is determined to be

Fig. 8.14 Variation of the
current with time

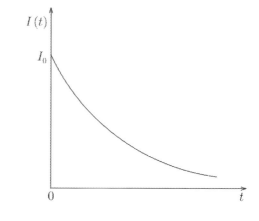

$$I(t) = \frac{E}{R}\exp\left(-\frac{t}{RC}\right).$$

The variation of the current with time is shown in Fig. 8.14.

$\diamond$

Example 8.3 Suppose that the power source is opened by turning switch S off at
$t = 0$ in a steady state at a sufficiently long time after turning on S in the circuit
shown in Fig. 8.9. Determine the current $I(t)$ for $t > 0$. We assume $R < 2(L/C)^{1/2}$.

Solution 8.3 The general solution (8.26) holds also in this case. In the initial state the
current does not flow, and Eq. (8.27) holds. This is because the first term in Eq. (8.21)
diverges if the current increases suddenly. The current is caused by a discharge from
the capacitor. Then, the equation is given by

$$\int_{0}^{\infty} I(t)\mathrm{d}t = -CE.$$

So, we have

$$\frac{K_1}{S_1} + \frac{K_2}{S_2} = CE,$$

where S_1 and S_2 are given by Eq. (8.25). The coefficients K_1 and K_2 are obtained as

$$K_1 = -K_2 = -\frac{E}{\left[R^2 - 4(L/C)\right]^{1/2}} = \frac{\mathrm{i}E}{2\alpha L}.$$

Thus, the current is determined to be

$$I(t) = -\frac{E}{\alpha L} \exp\left(-\frac{Rt}{2L}\right) \sin \alpha t.$$

◇

Example 8.4 Suppose that switch S is connected to the DC voltage source at $t = 0$ in the electric circuit shown in Fig. 8.15. Determined the current $I(t)$ for $t > 0$.

Fig. 8.15 Electric circuit composed of coil, resistor, and DC voltage source

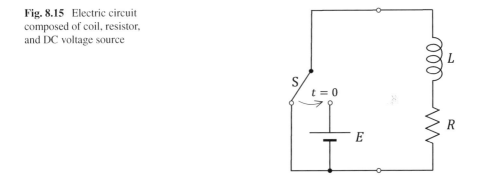

Solution 8.4 The equation for the circuit is written as

$$L\frac{dI}{dt} + RI = E.$$

Here, we put $I(t) = I_0 e^{St} + K$. Then, substituting this into the above equation, we have $S = -R/L$ and $K = E/R$. The initial condition, $I(t = 0) = 0$, yields $I_0 = -E/R$. Thus, the current is determined to be

$$I(t) = \frac{E}{R}\left[1 - \exp\left(-\frac{R}{L}t\right)\right].$$

◇

We have treated the transient response in the electric circuit shown in Fig. 8.9 after the DC voltage source is connected. Here, we determine the energy supplied by the power source, that consumed in the resistor, and those stored in the capacitor and coil.

For $R \neq 2(L/C)^{1/2}$, the current is given by Eqs. (8.25), (8.26), and (8.30). The energy supplied by the power source is

$$W = \int_0^\infty I(t)E\,dt = E\left[\frac{K_1}{S_1}e^{S_1 t} + \frac{K_2}{S_2}e^{S_2 t}\right]_0^\infty = CE^2. \tag{8.37}$$

The energy consumed in the resistor is

$$W_R = \int_0^\infty RI(t)^2\,dt = R\left[\frac{K_1^2}{2S_1}e^{2S_1 t} + \frac{2K_1 K_2}{S_1 + S_2}e^{(S_1 t + S_2 t)} + \frac{K_2^2}{2S_2}e^{2S_2 t}\right]_0^\infty$$

$$= -R\left(\frac{K_1^2}{2S_1} + \frac{2K_1 K_2}{S_1 + S_2} + \frac{K_2^2}{2S_2}\right). \tag{8.38}$$

Using the relationships

$$K_1^2 = K_2^2 = -K_1 K_2 = \frac{E^2}{R^2 - 4(L/C)} \tag{8.39}$$

and

$$S_1 + S_2 = -\frac{R}{L}, \quad S_1 S_2 = \frac{1}{LC}, \tag{8.40}$$

Equation (8.38) leads to

$$W_R = \frac{CE^2}{2}. \tag{8.41}$$

Since the electric charge $Q = CE$ is stored in the capacitor in the final state, the stored energy in the capacitor is obtained from Eq. (8.14) as

$$W_C = \frac{CE^2}{2}. \tag{8.42}$$

In the coil, the energy is stored initially, but is zero in the final state:

$$W_L = 0. \tag{8.43}$$

Thus, we have

$$W = W_R + W_C + W_L. \tag{8.44}$$

So, the law of conservation of energy holds. Some energy is stored in the capacitor, the same amount of energy is dissipated in the resistor, and no energy is stored in the coil. In the case of $R = 2(L/C)^{1/2}$, the same result is obtained by a different calculation.

Example 8.5 Determine the energy supplied by the DC voltage source, that dissipated in the resistor, and that stored in the capacitor in the transient process for the electric circuit treated in Example 8.2.

Solution 8.5 Using the solution for the current, the energy supplied by the power source is given by

$$W = \int_0^\infty I(t)E\,dt = \frac{E^2}{R}\int_0^\infty \exp\left(-\frac{t}{RC}\right)dt = CE^2,$$

and the energy dissipated by the resistor is

$$W_R = \int_0^\infty RI(t)^2\,dt = \frac{E^2}{R}\int_0^\infty \exp\left(-\frac{2t}{RC}\right)dt = \frac{1}{2}CE^2.$$

Since the energy stored in the capacitor is $W_C = CE^2/2$, the relation, $W = W_R + W_C$, holds, and the energy is conserved.

◇

Example 8.6 Suppose that switch S is turned from the DC voltage source to the capacitor at $t = 0$ in the electric circuit in the steady state in Fig. 8.16. Determine the current $I(t)$ for $t > 0$, the energy stored in each capacitor after a sufficiently long time, and the energy that has been dissipated in the resistor.

Fig. 8.16 Electric circuit composed of resistor, two capacitors, and DC voltage source

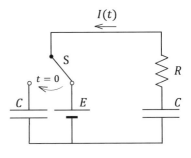

Solution 8.6 Since the voltage on the right capacitor is E in the initial condition, the electric charge and energy stored in it are CE and $U_1 = CE^2/2$, respectively.

After switch S is turned from the power source to the left capacitor, the current flows, the electric charge in the right capacitor decreases, and its voltage decreases as

$$V_{C1}(t) = E - \frac{1}{C} \int_0^t I(t)dt.$$

The electric charge is transferred to the left capacitor and its voltage changes as

$$V_{C2}(t) = \frac{1}{C} \int_0^t I(t)dt.$$

The voltage on the resistor is $V_R = RI(t)$. Then, the equation for this circuit is

$$\frac{2}{C} \int_0^t I(t)dt + RI(t) = E.$$

If we assume that the solution for the current is in the form of $I(t) = I_0 e^{St}$, the above equation leads to

$$\left(\frac{2}{CS} + R \right) I_0 e^{St} - \frac{2I_0}{CS} - E = 0.$$

Thus, we have

$$S = -\frac{2}{CR}, \quad I_0 = \frac{E}{R}.$$

The electric charge that has moved from the right capacitor to the left is

$$Q = I_0 \int_0^\infty e^{St} dt = \frac{CE}{2}.$$

Thus, the charge of the same amount, $CE/2$, remains in the right capacitor. The energy of $CE^2/8$ is stored in each capacitor. The energy dissipated in the resistor is

$$U_R = RI_0^2 \int_0^\infty e^{2St} dt = \frac{CE^2}{4}.$$

In conclusion, half of the energy stored in the right capacitor in the beginning is dissipated in the resistor, and a quarter of the energy is stored in each capacitor.

◇

8.3.2 Steady Response

Here, we treat the steady state response, in which the power source is an AC voltage source in the circuit shown in Fig. 8.7. We denote the electromotive force of the power source by $E \cos \omega t$. We assume that the solution for the current $I(t)$ is in the form:

$$I(t) = K_1 \cos \omega t + K_2 \sin \omega t. \tag{8.45}$$

Then, Eq. (8.17) becomes

$$\left[K_2 \left(\omega L - \frac{1}{\omega C} \right) + K_1 R \right] \cos \omega t - \left[K_1 \left(\omega L - \frac{1}{\omega C} \right) + K_2 R \right] \sin \omega t = E \cos \omega t. \tag{8.46}$$

Thus, we have

$$K_2 \left(\omega L - \frac{1}{\omega C} \right) + K_1 R = E, \tag{8.47}$$

$$K_1 \left(\omega L - \frac{1}{\omega C} \right) + K_2 R = 0. \tag{8.48}$$

The coefficients K_1 and K_2 are determined to be

$$K_1 = \frac{RE}{R^2 + [\omega L - 1/(\omega C)]^2}, \tag{8.49}$$

$$K_2 = \frac{[\omega L - 1/(\omega C)]E}{R^2 + [\omega L - 1/(\omega C)]^2}. \tag{8.50}$$

Then, the current is given by

$$\begin{aligned} I(t) &= \frac{E}{R^2 + [\omega L - 1/(\omega C)]^2} \left[R \cos \omega t + \left(\omega L - \frac{1}{\omega C} \right) \sin \omega t \right] \\ &= \frac{E}{Z_0} (\cos \theta \cos \omega t + \sin \theta \sin \omega t) = \frac{E}{Z_0} \cos(\omega t - \theta), \end{aligned} \tag{8.51}$$

where

$$Z_0 = \left[R^2 + \left(\omega L - \frac{1}{\omega C} \right)^2 \right]^{1/2},$$

(8.52)

$$\theta = \tan^{-1} \frac{1}{R} \left(\omega L - \frac{1}{\omega C} \right).$$

(8.53)

It can be understood that Z_0 corresponds to the resistance of the whole electric circuit and that the phase of the current is behind that of the voltage by θ. This type of analysis is much more easily done using the alternating current (AC) circuit theory, which will be introduced in Chap. 9.

Example 8.7 When AC voltage $E(t) = E \cos \omega t$ is applied to the circuit composed of a capacitor and coil shown in Fig. 8.17, determine the energies stored in the capacitor and coil. We assume that $\omega^2 LC \neq 1$.

Fig. 8.17 Electric circuit composed of coil, capacitor, and sinusoidal AC voltage source

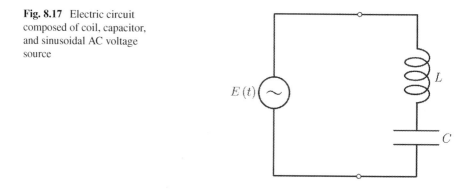

Solution 8.7 The equation for the circuit is given by

$$L \frac{dI(t)}{dt} + \frac{1}{C} \int I(t) dt = E \cos \omega t.$$

So, the current is obtained as

$$I(t) = \frac{\omega C E}{\omega^2 LC - 1} \sin \omega t.$$

Then, the energy stored in the coil is

$$W_L = \frac{1}{2} L I^2(t) = \frac{1}{2} L \left(\frac{\omega C E}{\omega^2 LC - 1} \right)^2 \sin^2 \omega t.$$

The electric charge stored in the capacitor is

$$Q(t) = -\frac{CE}{\omega^2 LC - 1} \cos \omega t$$

and the energy stored in the capacitor is determined to be

$$W_\text{C} = \frac{1}{2C} Q^2(t) = \frac{1}{2C} \left(\frac{CE}{\omega^2 LC - 1} \right)^2 \cos^2 \omega t.$$

Then, the sum of the energies stored in the coil and capacitor is

$$W_\text{L} + W_\text{C} = \frac{1}{2} \left(\frac{\omega CE}{\omega^2 LC - 1} \right)^2 \left[\left(L - \frac{1}{\omega^2 C} \right) \sin^2 \omega t + \frac{1}{\omega^2 C} \right]$$

$$= \frac{CE^2}{2(\omega^2 LC - 1)} \sin^2 \omega t + \frac{CE^2}{2(\omega^2 LC - 1)^2}.$$

On the other hand, the electric power supplied by the AC voltage source is

$$P(t) = E(t)I(t) = \frac{\omega CE}{\omega^2 LC - 1} \sin \omega t \cos \omega t.$$

So, the energy supplied by the source after $t = 0$ at which $I(t) = 0$ is

$$W(t) = \int_0^t P(t)dt = \frac{CE^2}{2(\omega^2 LC - 1)} \sin^2 \omega t.$$

Hence, the relation holds:

$$W(t) + \frac{CE^2}{2(\omega^2 LC - 1)^2} = W_\text{L} + W_\text{C}.$$

Thus, the energy moves between the source and the coil and capacitor, and is not dissipated. The second term on the left side in the above equation is written as $Q^2(0)/(2C)$ with the electric charge $Q(0)$ stored in the capacitor in the initial condition. So, it is the energy supplied by the power source that realizes the initial condition.

◇

Exercises

8.1 Determine the dissipated energy during one period ($0 \le t \le T$) when a sawtoothed-wave voltage shown in Fig. 8.18 is applied to a resistor with resistance R.

Fig. 8.18
Sawthoothed-wave voltage

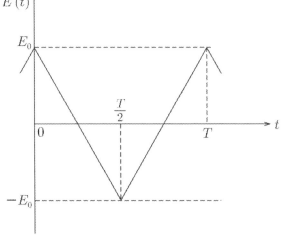

8.2 Suppose that the voltage on the circuit is changed from E to zero by turning switch S off at $t = 0$ in the electric circuit composed of a resistor and capacitor shown in Fig. 8.19. Determine the current for $t > 0$.

8.3 Solve the problem of Example 8.3 for $R > 2(L/C)^{1/2}$.

8.4 Solve the problem of Example 8.3 for $R = 2(L/C)^{1/2}$.

8.5 Suppose that the voltage on the circuit is changed to from E to zero by turning switch S off at $t = 0$ in the electric circuit composed of a coil and resistor shown in Fig. 8.20. Determine the current and the energy flow after the change.

8.6 Suppose the electric circuit composed of a resistor and capacitor in Exercise (2). Determine the energy dissipated in the resistor in the transient process.

Fig. 8.19 Electric circuit composed of resistor, capacitor, and DC voltage source

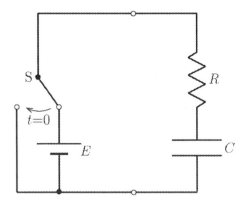

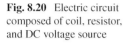

Fig. 8.20 Electric circuit composed of coil, resistor, and DC voltage source

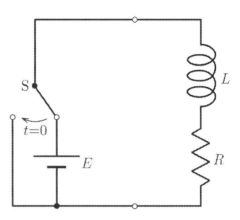

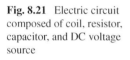

Fig. 8.21 Electric circuit composed of coil, resistor, capacitor, and DC voltage source

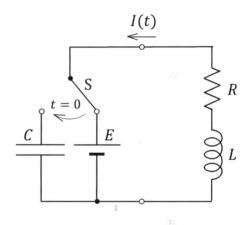

8.7 Suppose the electric circuit composed of a resistor and coil in Exercise (5). Determine the energy dissipated in the resistor for the transient process.

8.8 Determine the energy in each circuit element and that supplied by the AC power source for the steady process in the electric circuit composed of a coil, resistor, and capacitor treated in Sect. 8.3.2.

8.9 Suppose that switch is turned from the DC voltage source to the capacitor at $t = 0$ in the electric circuit shown in Fig. 8.21. Determine the current $I(t)$ for $t > 0$ and the energy dissipated in the resistor until a sufficiently long time has passed. Assume $R < 2(L/C)^{1/2}$.

8.10 Suppose that the output voltage of the AC voltage source is $E(t) = E_0 \cos \omega t$ in the electric circuit shown in Fig. 8.22. Determine the current.

Fig. 8.22 Electric circuit composed of capacitor, resistor, and AC voltage source

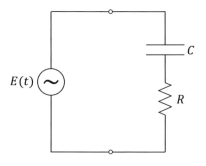

Chapter 9
Alternating Current Circuit

Abstract In this chapter we learn about alternating current (AC) circuit theory, i.e., the theoretical method in which the common time-dependent factor is abbreviated, for the case of application of sinusoidal AC voltage or current. The impedance is defined, which corresponds to the resistance in DC circuits, and we learn the relationship between the impedance and the energy for each circuit element. We treat various circuits composed of AC power source, resistor, capacitor, and coil, and learn the analysis method to determine the current. We also learn the characteristics of various types of filter from the frequency response. The advantage of using the complex electric power, which is composed of the effective electric power and reactive electric power, is introduced.

9.1 Sinusoidal Variation with Time

The voltage or current commonly changes with time in electric circuits, as discussed in Chap. 8. It is well known that an arbitrary variation with time can be expressed as a sum of sinusoidal alternating functions as

$$I(t) = \sum_{n=0}^{\infty} (A_n \cos n\omega t + B_n \sin n\omega t). \tag{9.1}$$

This is called the **Fourier series**. Constants A_n and B_n can be determined as follows:

$$A_n = \frac{1}{\pi} \int_0^{2\pi} I(t) \cos n\omega t \, \mathrm{d}\omega t; \quad n \geq 1, \tag{9.2}$$

$$A_0 = \frac{1}{\pi} \int_0^{2\pi} I(t) \mathrm{d}\omega t, \tag{9.3}$$

© The Author(s), under exclusive license to Springer Nature Switzerland AG 2023
T. Matsushita, *Electricity*,
https://doi.org/10.1007/978-3-031-44002-1_9

$$B_n = \frac{1}{\pi} \int_0^{2\pi} I(t) \sin n\,\omega t\,\mathrm{d}\omega t; \quad n \geq 1. \tag{9.4}$$

Hence, if we know the response to a sinusoidal AC voltage or current, any variation with time can be obtained by superimposing each response. Thus, understanding of the response to a sinusoidal variation is essential. For this reason, we treat here the cases in which the voltage or current varies sinusoidally with time.

Example 9.1 Expand the following sawtoothed-current wave, one period of which is given by

$$I(t) = I_0\left(1 - \frac{4t}{T}\right); \quad 0 \leq t \leq \frac{T}{2},$$

$$= I_0\left(-3 + \frac{4t}{T}\right); \quad \frac{T}{2} \leq t \leq T,$$

in a Fourier series with $\omega = 2\pi/T$ (see Fig. 9.1).

Fig. 9.1 Sawtoothed current wave

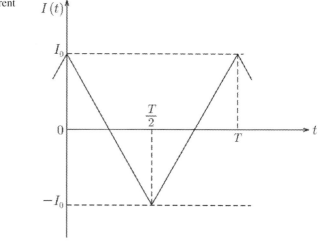

Solution 9.1 We derive easily $A_0 = 0$ using Eq. (9.3). If we put $\omega t = \phi$ for simplicity, we have

$$A_n = \frac{2I_0}{\pi^2}\left(-\int_0^{\pi} \phi \cos n\phi\,\mathrm{d}\phi + \int_{\pi}^{2\pi} \phi \cos n\phi\,\mathrm{d}\phi\right)$$

for $n \geq 1$. Using the relation

$$\int \phi \cos n\phi \, d\phi = \frac{\phi}{n} \sin n\phi + \frac{1}{n^2} \cos n\phi,$$

we have

$$A_n = \frac{4I_0}{n^2\pi^2}\left[1 - (-1)^n\right].$$

Similarly, we have

$$B_n = 0,$$

where the following relation was used:

$$\int \phi \sin n\phi \, d\phi = -\frac{\phi}{n} \cos n\phi + \frac{1}{n^2} \sin n\phi.$$

Using $A_{2m} = 0$ and $A_{2m+1} = 8I_0/\left[(2m+1)^2\pi^2\right]$, the current is expressed as

$$I(t) = \sum_{m=0}^{\infty} A_{2m+1} \cos(2m+1)\omega t = \sum_{m=0}^{\infty} \frac{8I_0}{(2m+1)^2\pi^2} \cos(2m+1)\omega t.$$

$\diamond$

9.2 Alternating Current Circuit Theory

We will now determine the current when a voltage source of the electromotive force $E_\mathrm{m} \cos \omega t$ is connected to a certain electric circuit, as treated in Sect. 8.3.2. If we use the complex relation,

$$e^{i\alpha} = \cos \alpha + i \sin \alpha, \tag{9.5}$$

the above electromotive force is equal to the real part of $E_\mathrm{m}e^{i\omega t}$:

$$E_\mathrm{m} \cos \omega t = \mathrm{Re}\, E_\mathrm{m}e^{i\omega t}, \tag{9.6}$$

where Re is a symbol indicating that we should take a real part. The electromotive force $E_\mathrm{m}e^{i\omega t}$ on the complex plane is shown in Fig. 9.2. It is a circle of radius E_m with the center on the origin, and the angle from the real axis is ωt, showing that the electromotive force circulates with the angular velocity ω in the counterclockwise direction. Such a notation, resembling a vector on the complex plane, is called **phasor**

Fig. 9.2 Phasor notation of
electromotive force on the
complex plane

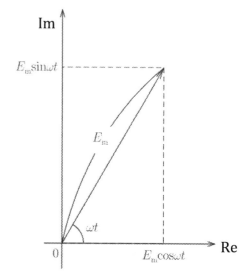

notation for distinction from a vector in real space. The analytic method using the
complex notation is **alternation current (AC) circuit theory**. Here, we show that
this method is useful for easily determining the solution.

We repeat the example treated in Sect. 8.3.2. Assume that the current is expressed
as

$$I_\mathrm{m}\mathrm{e}^{\mathrm{i}(\omega t+\varphi)}. \tag{9.7}$$

Then, its differentiation with time is

$$\frac{\mathrm{d}}{\mathrm{d}t}I(t) = \mathrm{i}\omega I_\mathrm{m}\mathrm{e}^{\mathrm{i}(\omega t+\varphi)} = \mathrm{i}\omega I(t), \tag{9.8}$$

which is a simple multiplication of $I(t)$ by $\mathrm{i}\omega$. The integral of the current with time
is

$$\int I(t)\mathrm{d}t = \frac{1}{\mathrm{i}\omega}I_\mathrm{m}\mathrm{e}^{\mathrm{i}(\omega t+\varphi)} = \frac{1}{\mathrm{i}\omega}I(t), \tag{9.9}$$

which is a simple division of $I(t)$ by $\mathrm{i}\omega$. Thus, Eq. (8.21) leads to

$$\left[R + \mathrm{i}\left(\omega L - \frac{1}{\omega C}\right)\right]I_\mathrm{m}\mathrm{e}^{\mathrm{i}\varphi} = E_\mathrm{m}, \tag{9.10}$$

where a common time-dependent factor $\mathrm{e}^{\mathrm{i}\omega t}$ is eliminated. The left side is written as
$Z_0 I_\mathrm{m}\mathrm{e}^{\mathrm{i}(\theta+\varphi)}$ using Eqs. (8.52) and (8.53). Thus, we have

$$I_{\mathrm{m}} = \frac{E_{\mathrm{m}}}{Z_0}, \tag{9.11}$$

$$\varphi = -\theta. \tag{9.12}$$

The real part of the current is given by

$$I(t) = \frac{E_{\mathrm{m}}}{Z_0} \cos(\omega t - \theta). \tag{9.13}$$

This agrees with the result obtained in Sect. 8.3.2. Thus, the analysis clearly becomes easier by the use of complex notation.

We use the real part of the complex notation to go back to the real phenomenon. It is also possible to use the imaginary part. The latter corresponds to the case in which the voltage of the source is given by $E_{\mathrm{m}} \sin \omega t$.

Since the complex notation holds when we analyze phenomena in AC circuits, it is possible to understand the phenomena directly from the complex notation without going back to the real part in each case. In addition, the common factor $\mathrm{e}^{\mathrm{i}\omega t}$ can be eliminated. In fact, the relationship between the current and the voltage can be obtained without this time-dependent factor, as is seen in Eq. (9.10). The obtained relationship between the current and the voltage can be understood as the relationship for peak values or effective values.

We treated the above example based on the condition that the initial phase of the voltage of the power source is zero. In the case of an electric circuit connected in series, the applied voltage is different for each element, while the current flowing through each element is common. So, it is more convenient to analyze based on the current, and hereafter we follow this style.

9.3 Impedance

When current $I\mathrm{e}^{\mathrm{i}\omega t}$ flows through a circuit element as shown in Fig. 9.3, the voltage on it is denoted by $V\mathrm{e}^{\mathrm{i}\omega t}$ and the relation between them is expressed as

$$V = ZI. \tag{9.14}$$

In this case Z is called the **impedance**. On the other hand, if the above relation is expressed as

$$I = YV, \tag{9.15}$$

where Y is called the **admittance**.

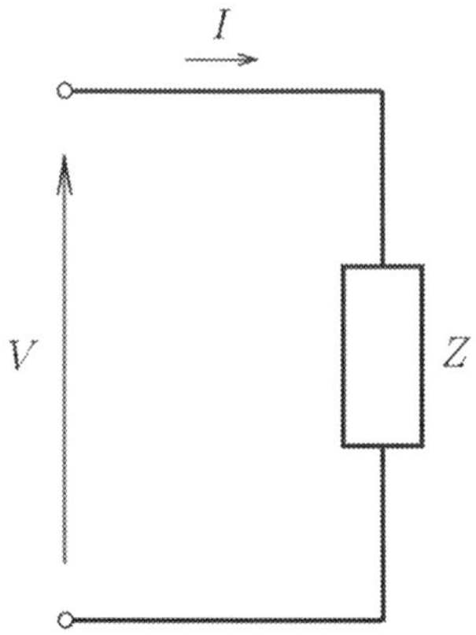

Fig. 9.3 Applied voltage
and current flowing through
a circuit element

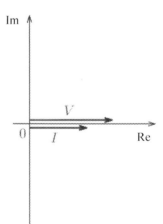

In the following we treat the impedance of each circuit element. There is also an impedance in the AC power source, which corresponds to the internal resistance in a DC power source, and this is called the **internal impedance**.

9.3.1 Resistor

From Eq. (8.1) the impedance of a resistor of resistance R is

$$Z = R. \tag{9.16}$$

The phasor notation of the voltage based on the flowing current for the resistor is shown in Fig. 9.4.

Fig. 9.4 Relation between
the current and voltage for a
resistor

9.3.2 Capacitor

The relation between the current and the voltage for a capacitor with capacity C is given from Eq. (8.5) as

$$V = \frac{I}{i\omega C}. \tag{9.17}$$

So, the impedance is

$$Z = \frac{1}{i\omega C} = -i\frac{1}{\omega C}. \tag{9.18}$$

The phasor notation of the voltage based on the flowing current for the capacitor is shown in Fig. 9.5. The phase of the voltage is behind that of the current by $\pi/2$.

Fig. 9.5 Relation between the current and voltage for a capacitor

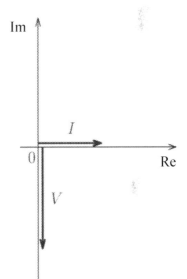

9.3.3 Coil

The relation between the current and the voltage for a coil with inductance L is given from Eq. (8.8) as

$$V = i\omega L I.   \tag{9.19}$$

So, the impedance is

$$Z = i\omega L.   \tag{9.20}$$

The phasor notation of the voltage based on the flowing current for the coil is shown in Fig. 9.6. The phase of the voltage is ahead that of the current by $\pi/2$.

Fig. 9.6 Relation between the current and voltage for a coil

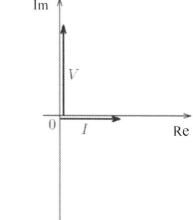

9.4   Series Connection and Parallel Connection

When circuit elements of impedances $Z_i (i = 1, 2, \ldots, n)$ are connected in series, the combined impedance is similarly given as the combined resistance as

$$Z = \sum_{i=1}^{n} Z_i.   \tag{9.21}$$

Hence, using the admittance of each circuit component, $Y_i = 1/Z_i (i = 1, 2, \ldots, n)$, the combined admittance, $Y = 1/Z$, is given by

$$\frac{1}{Y} = \sum_{i=1}^{n} \frac{1}{Y_i}.   \tag{9.22}$$

When these circuit elements are connected in parallel, the combined impedance is similarly given as the combined resistance

$$\frac{1}{Z} = \sum_{i=1}^{n} \frac{1}{Z_i}, \tag{9.23}$$

and the combined admittance is

$$Y = \sum_{i=1}^{n} Y_i. \tag{9.24}$$

Here, we will determine the combined impedance and admittance for the connected circuit elements shown in Fig. 9.7. Since the coil is connected in series with the resistor and capacitor, which are connected in parallel, the combined impedance is

$$Z = i\omega L + \frac{R(i\omega C)^{-1}}{R + (i\omega C)^{-1}} = i\omega L + \frac{R(1 - i\omega C R)}{1 + (\omega C R)^2}$$

$$= \frac{R}{1 + (\omega C R)^2} + i\left[\omega L - \frac{\omega C R^2}{1 + (\omega C R)^2}\right]. \tag{9.25}$$

It is desirable to describe such an expression by a divided form of real and imaginary parts.

Then, we determine the combined admittance. Since elements of admittance $1/i\omega L$ and $R^{-1} + i\omega C$ are connected in series, the combined admittance is

$$Y = \frac{(i\omega L)^{-1}\left(R^{-1} + i\omega C\right)}{(i\omega L)^{-1} + R^{-1} + i\omega C} = \frac{1 + i\omega C R}{R\left(1 - \omega^2 L C\right) + i\omega L}$$

$$= \frac{R + i\omega\left[C R^2\left(1 - \omega^2 L C\right)^2 - L\right]}{R^2\left(1 - \omega^2 L C\right)^2 + \omega^2 L^2}. \tag{9.26}$$

This can also be obtained from $Y = Z^{-1}$ with the result of Eq. (9.25).

Fig. 9.7 Connected circuit elements

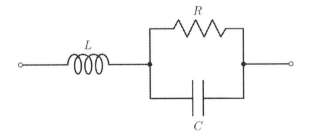

As shown above, the impedance is generally a complex number. Thus, the impedance is written in a separate form of real and imaginary parts as

$$Z = R + iX, \tag{9.27}$$

where R and X are called the **resistance** (or **resistance component**) and the **reactance**, respectively. The admittance is also a complex number represented by

$$Y = G + iB, \tag{9.28}$$

where G and B are called the **conductance** and the **susceptance**, respectively. From the relation, $ZY = 1$, we have

$$G = \frac{R}{R^2 + X^2}, \tag{9.29}$$

$$B = -\frac{X}{R^2 + X^2}. \tag{9.30}$$

R and X are similarly expressed using G and B (see Exercise 10.3).

Thus, the reactance X has a positive or negative value: It is called the **inductive reactance** for $X > 0$, and the **capacitive reactance** for $X < 0$.

Example 9.2 Determine the impedance, admittance, resistance, reactance, conductance, and susceptance for the connected circuit elements shown in Fig. 9.8.

Fig. 9.8 Connected circuit elements

Solution 9.2 Since the resistor is connected in series with the capacitor and coil which are connected in parallel, the impedance is given by

$$Z = R_1 + \frac{L/C}{i\omega L + (i\omega C)^{-1}} = R_1 + \frac{i\omega L}{1 - \omega^2 LC},$$

and the resistance and reactance are, respectively, given by

$$R = R_1, \quad X = \frac{\omega L}{1 - \omega^2 LC}.$$

Since the elements of admittance $1/R_1$ and $i\omega C + (i\omega L)^{-1}$ are connected in series, the combined admittance is

$$Y = \frac{R_1^{-1}\left[i\omega C + (i\omega L)^{-1}\right]}{R_1^{-1} + i\omega C + (i\omega L)^{-1}} = \frac{1 - \omega^2 LC}{R_1(1 - \omega^2 LC) + i\omega L}$$

$$= \frac{R_1(1 - \omega^2 LC)^2 - i\omega L(1 - \omega^2 LC)}{R_1^2(1 - \omega^2 LC)^2 + \omega^2 L^2}.$$

The conductance and susceptance are, respectively, given by

$$G = \frac{R_1(1 - \omega^2 LC)^2}{R_1^2(1 - \omega^2 LC)^2 + \omega^2 L^2}, \quad B = -\frac{\omega L(1 - \omega^2 LC)}{R_1^2(1 - \omega^2 LC)^2 + \omega^2 L^2}.$$

$\diamondsuit$

9.5 Alternating Current Circuit

We learned about circuits composed of resistors and DC power sources in Chap. 7. We can similarly analyze the response in alternating current (AC) circuits composed of coils, resistors, capacitors, and AC power sources. In this case resistances are merely changed to impedances. For example, assume the electric circuit shown in Fig. 9.9. The impedance of each circuit element is denoted by Z_1, Z_2, Z_3, and Z_4. If these are replaced by resistances R, R_2, R_3, and R_4, and an AC voltage source is replaced by a DC voltage source, this reduces to the resistor circuit shown in Fig. 7.14. So, we will repeat the corresponding analysis here.

Since the current that flows from A to D through B is

$$I_1 = \frac{E}{Z_1 + Z_2}, \tag{9.31}$$

the voltage on Z_1 and Z_2 are, respectively, given by

$$V_1 = Z_1 I_1 = \frac{Z_1 E}{Z_1 + Z_2}, \quad V_2 = Z_2 I_1 = \frac{Z_2 E}{Z_1 + Z_2}. \tag{9.32}$$

The current flowing from A to D through C is similarly given by

$$I_3 = \frac{E}{Z_3 + Z_4}, \tag{9.33}$$

Fig. 9.9 AC bridge

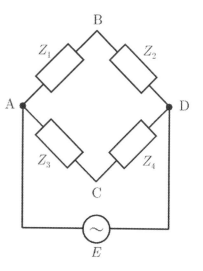

and the voltages on Z_3 and Z_4 are, respectively, given by

$$V_3 = Z_3 I_3 = \frac{Z_3 E}{Z_3 + Z_4}, \quad V_4 = Z_4 I_3 = \frac{Z_4 E}{Z_3 + Z_4}. \tag{9.34}$$

Using the above results, we determine the voltage between points B and C, i.e., the electric potential of point B measured from point C. The voltage between points D and C is $-V_4$ and that between points B and D is V_2. So, the voltage between points B and C is

$$
\begin{aligned}
V_{\mathrm{BC}} = V_2 - V_4 &= \frac{[Z_2(Z_3 + Z_4) - Z_4(Z_1 + Z_2)]E}{(Z_1 + Z_2)(Z_3 + Z_4)} \\
&= \frac{(Z_2 Z_3 - Z_4 Z_1)E}{(Z_1 + Z_2)(Z_3 + Z_4)}.
\end{aligned} \tag{9.35}
$$

This result is the same as that in Eq. (7.20), if each resistance is replaced by corresponding impedance. So, the condition that the voltage between points B and C becomes zero is given by

$$Z_2 Z_3 = Z_4 Z_1. \tag{9.36}$$

The circuit shown in Fig. 9.9 is called the **AC bridge**.

In the case of the electric circuit shown in Fig. 9.10, the following relation must be satisfied so that the voltage between points B and C is zero:

$$\frac{L_3}{C_2} = R_4 R_1. \tag{9.37}$$

Fig. 9.10 Example of AC bridge

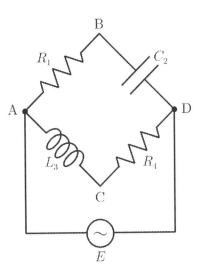

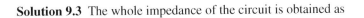

Example 9.3 Determine voltage V_1 on the resistor R_1 in the circuit shown in Fig. 9.11.

Fig. 9.11 AC circuit

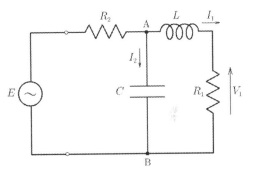

Solution 9.3 The whole impedance of the circuit is obtained as

$$Z = R_2 + \frac{(i\omega C)^{-1}(R_1 + i\omega L)}{(i\omega C)^{-1} + R_1 + i\omega L} = R_2 + \frac{R_1 + i\omega L}{1 + i\omega C(R_1 + i\omega L)}.$$

Hence, the whole current is given by $I = E/Z$. If we denote by I_1 and I_2 the current flowing through the coil and resistor and that flowing through the capacitor, respectively, as in the figure, we have $I = I_1 + I_2$. Another condition is that the voltage between points A and B is given by

$$(R_1 + i\omega L)I_1 = \frac{I_2}{i\omega C}.$$

From these conditions we have

$$I_1 = \frac{I}{i\omega C(R_1 + i\omega L) + 1} = \frac{E}{R_1 + R_2(1 - \omega^2 LC) + i\omega(L + CR_1 R_2)}.$$

Thus, the voltage on the resistor is determined to be

$$V_1 = R_1 I_1 = \frac{R_1 E}{R_1 + R_2(1 - \omega^2 LC) + i\omega(L + CR_1 R_2)}.$$

$\diamond$

Example 9.4 Derive the condition so that the phase of I_2 is behind that of E by $\pi/2$ in the circuit shown in Fig. 9.12.

Fig. 9.12 AC circuit

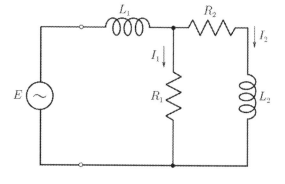

Solution 9.4 The impedance of the whole circuit is

$$Z = i\omega L_1 + \frac{R_1(R_2 + i\omega L_2)}{R_1 + R_2 + i\omega L_2}$$

and the whole current is $I = E/Z$. If we denote the current flowing through R_1 by I_1, we have $I_1 + I_2 = I$ and

$$R_1 I_1 = (R_2 + i\omega L_2)I_2.$$

From these conditions, I_2 is obtained:

$$I_2 = \frac{E}{Z} \cdot \frac{R_1}{R_1 + R_2 + i\omega L_2} = \frac{R_1 E}{R_1 R_2 - \omega^2 L_1 L_2 + i\omega[(R_1 + R_2)L_1 + R_1 L_2]}.$$

The requirement is satisfied when this is a negative imaginary number. Thus, the required condition is

$$R_1 R_2 - \omega^2 L_1 L_2 = 0.$$

◇

9.6 Frequency Response

There are many cases in which variation with time is not simply sinusoidal or not steady. Even in such cases, any variation with time can be realized by superimposing sinusoidal variations with different frequencies, as stated in Sect. 9.1. So, it is important to understand the difference in the response of AC circuits for different frequencies. We will treat the frequency response of various circuits in this section.

We first compare the frequency response of each circuit component. The impedance of the resistor is given by Eq. (9.16) and is unchanged when the angular frequency ω changes. The impedance of the capacitor is given by Eq. (9.18) and its magnitude becomes small when ω increases. On the other hand, the impedance of the coil is given by Eq. (9.20) and its magnitude becomes large when ω increases. So, electric circuits composed of these components may have complicated dependencies on ω.

Here, we treat the frequency response for the circuit composed of a coil, resistor, and capacitor, as shown in Fig. 9.13. The impedance of this circuit is

$$Z = R + i\left(\omega L - \frac{1}{\omega C}\right). \tag{9.38}$$

From Eq. (9.10) the current flowing through the circuit is written as

Fig. 9.13 Series resonance circuit

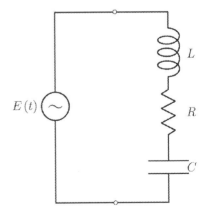

$$I = \frac{|E|}{|Z|}e^{i\varphi} \tag{9.39}$$

with reference to the electromotive force of the power source, where

$$|Z| = \left[R^2 + \left(\omega L - \frac{1}{\omega C} \right)^2 \right]^{1/2}, \tag{9.40}$$

$$\varphi = -\tan^{-1}\frac{1}{R}\left(\omega L - \frac{1}{\omega C} \right). \tag{9.41}$$

We suppose that the angular frequency becomes very small. In this case $|Z|$ is approximately given by $1/\omega C$, and the magnitude of the current is approximately given by

$$|I| \simeq \omega C |E|. \tag{9.42}$$

Thus, it becomes very small and its phase is approximately equal to $\pi/2$. On the contrary, if ω becomes very large, $|Z|$ is approximately equal to ωL, the magnitude of the current is very small as

$$|I| \simeq \frac{|E|}{\omega L}, \tag{9.43}$$

and its phase is approximately equal to $-\pi/2$. In the middle area of ω, the current takes on a large value and its phase takes on a middle value. Most notably, the reactance is close to zero. The current takes on its maximum value, $|I| = |E|/R$, and its phase is zero at

$$\omega = \frac{1}{(LC)^{1/2}} \equiv \omega_0. \tag{9.44}$$

The value of ω_0 given by Eq. (9.44) is called the **resonance angular frequency** and the electric circuit shown in Fig. 9.13 is called a **series resonance circuit**. Here, we put

$$Q_f = \frac{1}{R}\left(\frac{L}{C} \right)^{1/2}, \tag{9.45}$$

and then, we have

$$\left(\frac{L}{C} \right)^{1/2}\frac{|I|}{|E|} = Q_f\left[1 + Q_f^2\left(\frac{\omega}{\omega_0} - \frac{\omega_0}{\omega} \right)^2 \right]^{-1/2}, \tag{9.46}$$

$$\varphi = -\tan^{-1} Q_f \left(\frac{\omega}{\omega_0} - \frac{\omega_0}{\omega} \right). \tag{9.47}$$

The angular frequency dependences of the magnitude and phase of the current are shown in Figs. 9.14 and 9.15, respectively. The sharpness of the resonance depends strongly on the value of Q_f. Q_f is called the **quality factor**, or **Q value**.

Suppose that $Q_f \gg 1$. If the value of $|I|$ reduces to $1/\sqrt{2}$ of its peak value at $\omega = \omega_0 + \Delta\omega$, we have

$$\frac{1}{Q_f} = \frac{\omega_0 + \Delta\omega}{\omega_0} - \frac{\omega_0}{\omega_0 + \Delta\omega} \simeq \frac{2\Delta\omega}{\omega_0}. \tag{9.48}$$

$|I|$ reduces also approximately to $1/\sqrt{2}$ of its peak value at $\omega = \omega_0 - \Delta\omega$. Since the power that can be taken out the power source is proportional to $|I|^2$, the power takes on one half of its maximum value at $\omega_0 \pm \Delta\omega$. The band width $2\Delta\omega$ in the angular frequency range is called the **full width at half maximum** and is approximately given by

$$2\Delta\omega \simeq \frac{\omega_0}{Q_f}. \tag{9.49}$$

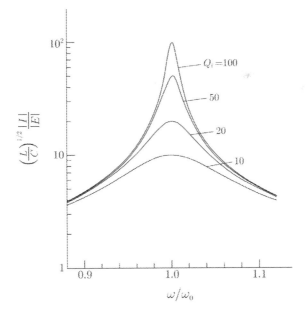

Fig. 9.14 Dependence of the magnitude of the current in a series resonance circuit on angular frequency for $Q_f = 10, 20, 50,$ and 100

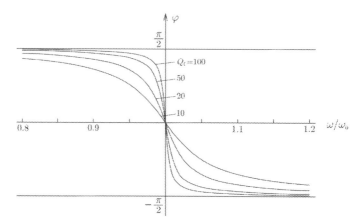

Fig. 9.15 Dependence of the phase of the current in a series resonance circuit on angular frequency for $Q_f = 10, 20, 50,$ and 100

Example 9.5 Derive the magnitude and the angular frequency dependence of the current I in the circuit shown in Fig. 9.16.

Fig. 9.16 Parallel resonance circuit

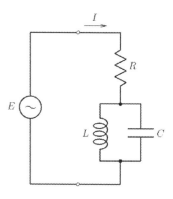

Solution 9.5 The impedance of this circuit was derived in Example 9.2 as

$$Z = R + \frac{i\omega L}{1 - \omega^2 LC}.$$

Hence, the magnitude and the phase of the current measured from the electromotive force are, respectively, given by

$$|I| = |E|\left[R^2 + \left(\frac{\omega L}{1 - \omega^2 LC}\right)^2\right]^{-1/2},$$

$$\varphi = -\tan^{-1}\frac{1}{R}\left(\frac{\omega L}{1 - \omega^2 LC}\right).$$

◇

When ω is very small, the magnitude of the current is approximately constant and takes on a value of about $|E|/R$, and its phase is approximately zero. When ω is very large, the magnitude of the current is again approximately constant and takes on a value of about $|E|/R$, and its phase is approximately zero. The value of the current becomes zero at the **resonance angular frequency** given by

$$\omega = \frac{1}{(LC)^{1/2}} \equiv \omega_0.$$

The phase is $-\pi/2$ at ω just smaller than ω_0, and is $\pi/2$ at ω just larger than ω_0. These dependencies on ω are shown in Figs. 9.17 and 9.18. Although the total current is zero at $\omega = \omega_0$, the electromotive force is fully applied to the coil and capacitance, and the circulating current I' flows as illustrated in Fig. 9.19. This current satisfies

$$LI' = \frac{E}{i\omega_0} = -i\omega_0 CE.$$

This circuit is called a **parallel resonance circuit**. The case in which a resistor is also connected in parallel is included.

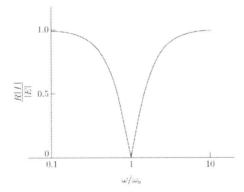

Fig. 9.17 Angular frequency dependence of the magnitude of the current in a parallel resonance circuit for $R = \omega_0 L/\sqrt{2}$

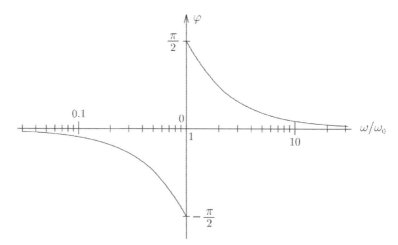

Fig. 9.18 Angular frequency dependence of the phase of the current in a parallel resonance circuit for $R = \omega_0 L / \sqrt{2}$

Fig. 9.19 The current flowing in a parallel resonance circuit at $\omega = \omega_0$

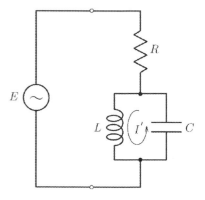

9.7   Filter

Here, we determine the voltage V_2 on the resistor (R) in the circuit shown in Fig. 9.20. Since the impedance of the resistor and capacitor connected in parallel is

$$Z = \frac{R}{1 + i\omega C R},$$

(9.50)

we have

$$\frac{V_2}{V_1} = \frac{Z}{Z + i\omega L} = \frac{R}{R(1 - \omega^2 LC) + i\omega L}.$$

(9.51)

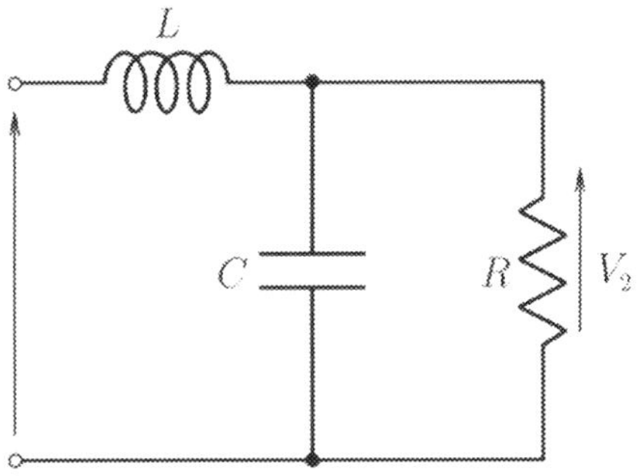

Fig. 9.20 Low-pass filter circuit

$$L$$

$$V_1 \qquad C \qquad R \quad V_2$$

Using the factor given by

$$\omega_c = \frac{1}{(LC)^{1/2}}, \tag{9.52}$$

Equation (9.51) leads to

$$\left| \frac{V_2}{V_1} \right| = \left[\left(1 - \frac{\omega^2}{\omega_c^2} \right)^2 + \left(\frac{\omega L}{R} \right)^2 \right]^{-1/2}. \tag{9.53}$$

For the resistance satisfying

$$R = \frac{\omega_c L}{\sqrt{2}}, \tag{9.54}$$

Equation (9.53) is reduced to

$$\left| \frac{V_2}{V_1} \right| = \left[1 + \left(\frac{\omega}{\omega_c} \right)^4 \right]^{-1/2}. \tag{9.55}$$

Thus, the right side is almost equal to 1 for $\omega \ll \omega_c$. On the other hand, for $\omega \gg \omega_c$, the right side is almost given by $(\omega_c/\omega)^2$ and decreases with increasing ω. This angular frequency dependence is shown in Fig. 9.21. Such a circuit that passes only a signal of frequency lower than ω_c and cuts that higher than ω_c is called a **low-pass filter** and ω_c is called the **cut-off angular frequency**. So, $f_c = \omega_c/2\pi$ is the **cut-off frequency**.

The reason why such a characteristic is obtained is that the voltage on the resistor becomes lower due to the increase in the impedance of the coil in addition to the shortage of resistance due to the decrease in the impedance of the capacitor at high angular frequencies.

Next, we treat the electric circuit shown in Fig. 9.22. The impedance of the resistor and coil connected in parallel is

$$Z = \frac{i\omega L R}{R + i\omega L} = \frac{\omega L R}{\omega L - iR}. \tag{9.56}$$

Fig. 9.21 Angular
frequency characteristic of
low-pass filter

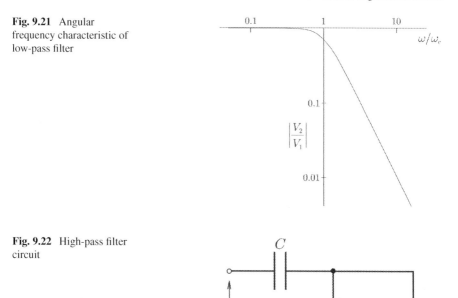

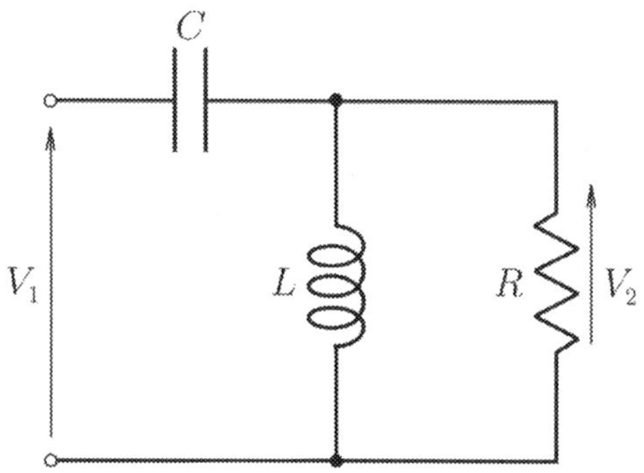

Fig. 9.22 High-pass filter
circuit

So, we have

$$\frac{V_2}{V_1} = \frac{Z}{Z + (i\omega C)^{-1}} = -\frac{\omega^2 LCR}{R(1 - \omega^2 LC) + i\omega L}.$$ (9.57)

Using Eqs. (9.52) and (9.54), Eq. (9.57) is reduced to

$$\left|\frac{V_2}{V_1}\right| = \left[1 + \left(\frac{\omega_c}{\omega}\right)^4\right]^{-1/2}.$$ (9.58)

For $\omega \ll \omega_c$, the right side is approximately equal to $(\omega/\omega_c)^2$ and becomes almost zero for small ω. For $\omega \gg \omega_c$, the right side approaches 1. The obtained angular frequency dependence is shown in Fig. 9.23. Such a circuit that passes only a signal of angular frequency higher than the designated value and cuts a signal with a lower angular frequency is called a **high-pass filter**. ω_c in this case is also called the cut-off angular frequency.

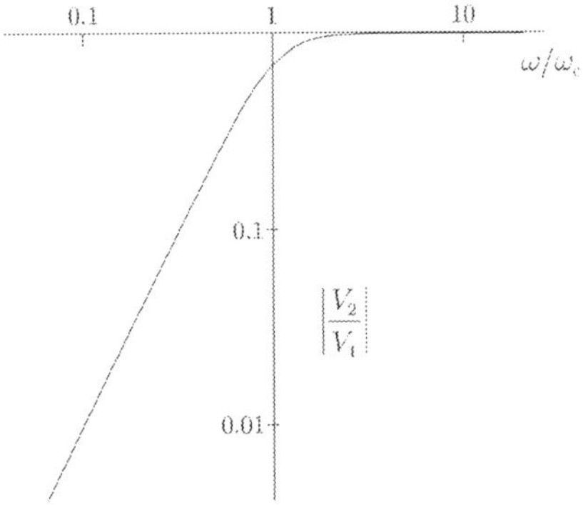

Fig. 9.23 Angular frequency characteristic of high-pass filter

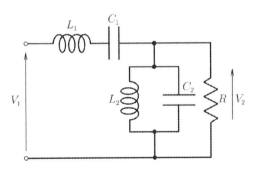

The reason why such a characteristic is obtained is that the voltage on the resistor becomes lower due to the increase in the impedance of the capacitor in addition to the shortage of resistance due to the decrease in the impedance of the coil at low angular frequencies.

Example 9.6 The circuit shown in Fig. 9.24 is a **band-pass filter** that passes only signals within a designated range of angular frequency. This circuit is designed so that the voltage on the resistor is lowered when the angular frequency is higher or lower. To realize this condition, it is necessary to adjust the capacitance of the capacitor. Determine the angular-frequency dependence of $|V_2/V_1|$ under the conditions:

$$L_1 = \frac{R}{\Delta\omega}, \quad L_2 = \frac{\Delta\omega R}{\omega_c^2}, \quad C_1 = \frac{\Delta\omega}{\omega_c^2 R}, \quad C_2 = \frac{1}{\Delta\omega R}.$$

Fig. 9.24 Band-pass filter circuit

Solution 9.6 Similar calculation to that above leads to

$$\frac{V_2}{V_1} = \left[1 + \frac{L_1}{L_2} + \frac{C_2}{C_1} - \omega^2 L_1 C_2 - \frac{1}{\omega^2 L_2 C_1} + i\frac{1}{R}\left(\omega L_1 - \frac{1}{\omega C_1}\right) \right]^{-1}.$$

Substituting the above conditions, we have

$$\left|\frac{V_2}{V_1}\right| = \left[\frac{1}{\Delta\omega^4}\left(\omega - \frac{\omega_c^2}{\omega}\right)^4 - \frac{1}{\Delta\omega^2}\left(\omega - \frac{\omega_c^2}{\omega}\right)^2 + 1\right]^{-1/2}$$

It is found that $|V_2/V_1|$ is equal to 1 at $\omega = \omega_c$ and reduces rapidly when the angular frequency deviates from ω_c. For example, $|V_2/V_1|$ takes on a value $1/\sqrt{3}$ at $\omega = \omega_c \pm \left(1/\sqrt{2}\right)\Delta\omega$, where $\Delta\omega$ is much smaller than ω_c. Thus, ω_c and $\Delta\omega$ represent the mean angular frequency and the band width, respectively. An angular frequency characteristic is shown in Fig. 9.25.

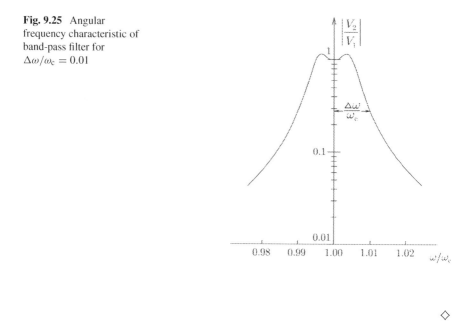

Fig. 9.25 Angular frequency characteristic of band-pass filter for $\Delta\omega/\omega_c = 0.01$

9.8   Alternating Current Power

When AC voltage is applied to a resistor, the electric power is given by Eq. (8.12), and the electric power is dissipated as discussed in Example 8.1. This is shown by the fact that the power is always positive and is continuously supplied by the power source, while the situation does not change. On the other hand, the electric power is not dissipated in capacitors or coils and goes back and forth between these elements and the power source, as shown in Sects. 8.2.2 and 8.2.3.

This can be shown by the fact that the power takes on a positive or negative value depending on the situation. The applied voltage and flowing current have the same directions in the phasor notation for the resistor. But those are perpendicular to each other for the coil and the capacitor, as shown in Figs. 9.5 and 9.6.

It should be noted that the phasor natation cannot be used for the electric power, since its variation with time contains a sinusoidal term of angular frequency 2ω. For this reason, it is necessary to treat practical variation with time for the electric power.

When I is a phasor notation of the current that flows through a load with the impedance and admittance given by Eqs. (9.27) and (9.28), respectively, the phasor notation of V is

$$V = ZI = (R + iX)I = |Z||I|e^{i\theta}, \tag{9.59}$$

where θ is given by

$$\cos\theta = \frac{R}{\left(R^2 + X^2\right)^{1/2}} = \frac{G}{\left(G^2 + B^2\right)^{1/2}}. \tag{9.60}$$

Then, the instantaneous values of the current and voltage on the load are, respectively, given by

$$I_i(t) = \sqrt{2}|I|\cos\omega t, \tag{9.61}$$

$$V_i(t) = \sqrt{2}|V|\cos(\omega t + \theta), \tag{9.62}$$

where $|V| = |Z||I|$. So, the instantaneous value of the electric power given to the load is

$$\begin{aligned}
P_i(t) &= V_i(t)I_i(t) = 2|V||I|\cos\omega t \cos(\omega t + \theta) \\
&= |V||I|\cos\theta(1 + \cos 2\omega t) - |V||I|\sin\theta \sin 2\omega t.
\end{aligned} \tag{9.63}$$

The first term is

$$R I_i^2(t), \tag{9.64}$$

which gives the electric power dissipated in the resistor. When this is averaged with respect to time over a period, $T = 2\pi/\omega$, we have

$$P = |V||I|\cos\theta = R|I|^2 = G|V|^2. \tag{9.65}$$

This is called the **effective power**, and $\cos\theta$ is the **power factor**. The second term in Eq. (9.63) is a power that goes back and forth between the load and the power source, and its average with respect to time is zero. The value that corresponds to its amplitude,

$$P_r = -|V||I|\sin\theta = -X|I|^2 = B|V|^2, \tag{9.66}$$

is called the **reactive power**, and the quantity common to both the effective and reactive powers,

$$P_a = |V||I|, \tag{9.67}$$

is called the **apparent power**. Although the unit of the effective power is [W], which is the same as the instantaneous power, that of the reactive power is [Var], which is abbreviation of Volt Ampere Reactive, and that of the apparent power is [VA].

When the voltage and current are V and I for a circuit or a load, it is convenient to use the **complex power** defined below to express the power:

$$
\begin{aligned}
P_e &= V^*I = |V||I|[\cos(\omega t + \theta) - i\sin(\omega t + \theta)](\cos \omega t + i\sin \omega t) \\
&= |V||I|\cos \theta - i|V||I|\sin \theta = P + iP_r,
\end{aligned} \tag{9.68}
$$

where the symbol * represents a complex conjugate, and P and P_r are the effective and reactive powers, respectively. So, we have

$$P = \mathrm{Re}(P_e) = \frac{1}{2}(P_e + P_e^*), \tag{9.69}$$

$$P_r = \mathrm{Im}(P_e) = \frac{1}{2i}(P_e - P_e^*), \tag{9.70}$$

where Im represents an imaginary part.

Here, we determine the dissipated power in the resistance in the circuit shown in Fig. 9.26. The current flowing through the resistor is

$$I' = \frac{L_2 V}{(L_1 + L_2)R + i\omega L_1 L_2}. \tag{9.71}$$

Then, the dissipated power is.

Fig. 9.26 Electric circuit

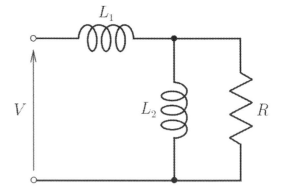

$$P = R|I'|^2 = \frac{RL_2^2|V|^2}{(L_1 + L_2)^2 R^2 + (\omega L_1 L_2)^2}. \tag{9.72}$$

Since the power is dissipated only in the resistor in this circuit, it can be determined as a power dissipated in the circuit. The current in the whole circuit is

$$I = \frac{V(R + i\omega L_2)}{-\omega^2 L_1 L_2 + i\omega(L_1 + L_2)R} \tag{9.73}$$

and the whole complex power is

$$P_e = V^*I = \frac{(R + i\omega L_2)|V|^2}{-\omega^2 L_1 L_2 + i\omega(L_1 + L_2)R}$$
$$= \frac{R\omega L_2^2 - i[(L_1 + L_2)R^2 + \omega^2 L_1 L_2]}{\omega[(L_1 + L_2)^2 R^2 + (\omega L_1 L_2)^2]}|V|^2. \tag{9.74}$$

From the real part, the dissipated power is determined to be

$$P = \frac{RL_2^2|V|^2}{(L_1 + L_2)^2 R^2 + (\omega L_1 L_2)^2}, \tag{9.75}$$

which agrees with the above result.

Example 9.7 Determine the total electric power dissipated in the circuit shown in Fig. 9.27 using the following methods:

(a) Add the power dissipated in each resistor,
(b) Determine the total power from the complex power.

Fig. 9.27 Electric circuit with two resistors

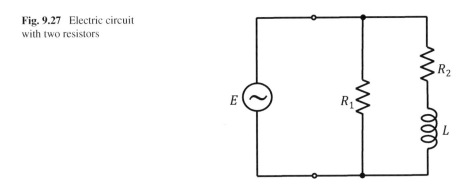

Solution 9.7

(a) The current flowing through resistance R_1 is $I_1 = E/R_1$, and the dissipated power is

$$P_1 = R_1|I_1|^2 = \frac{|E|^2}{R_1}.$$

The current flowing through resistance R_2 is $I_2 = E/(R_2 + i\omega L)$, and the dissipated power is

$$P_2 = R_2|I_2|^2 = \frac{R_2|E|^2}{R_2^2 + (\omega L)^2}.$$

Thus, the total dissipated power is determined to be

$$P = P_1 + P_2 = \frac{\left[(R_1 + R_2)R_2 + (\omega L)^2\right]|E|^2}{R_1\left[R_2^2 + (\omega L)^2\right]}.$$

(b) The total current of the circuit is

$$I = \frac{(R_1 + R_2 + i\omega L)E}{R_1(R_2 + i\omega L)},$$

and the complex power of the circuit is

$$P_e = E^*I = \frac{(R_1 + R_2 + i\omega L)(R_2 - i\omega L)|E|^2}{R_1\left[R_2^2 + (\omega L)^2\right]}.$$

Thus, the dissipated power is given by

$$\mathrm{Re}(P_e) = \frac{\left[(R_1 + R_2)R_2 + (\omega L)^2\right]|E|^2}{R_1\left[R_2^2 + (\omega L)^2\right]},$$

which agrees with the sum of the individual powers.

$\diamondsuit$

Example 9.8 Suppose that a load of impedance $Z = R + iX$ is connected to a power source of electromotive force E and internal impedance $Z_0 = R_0 + iX_0$, as shown in Fig. 9.28. Determine the condition in which the power dissipated in the load has its maximum value.

Fig. 9.28 AC power source with internal impedance and load

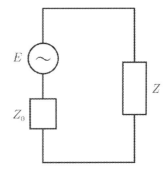

Solution 9.8 The voltage on the load is

$$V = \frac{(R + iX)E}{R + R_0 + i(X + X_0)},$$

and the current flowing through the load is

$$I = \frac{E}{R + R_0 + i(X + X_0)}.$$

Then, the complex power applied to the load is given by

$$P_e = V^*I = \frac{(R - iX)E^*}{R + R_0 - i(X + X_0)} \cdot \frac{E}{R + R_0 + i(X + X_0)}$$
$$= \frac{(R - iX)|E|^2}{(R + R_0)^2 + (X + X_0)^2}.$$

and the dissipated power is determined to be

$$P = \text{Re}(P_e) = \frac{R|E|^2}{(R + R_0)^2 + (X + X_0)^2}.$$

It is easily found that P takes on its maximum value, $P = R|E|^2/(R + R_0)^2$, when

$$X = -X_0 \tag{9.76}$$

is satisfied under variation of X. To obtain the maximum P under variation of R, from the condition $\partial P/\partial R = 0$, we have

$$R = R_0. \tag{9.77}$$

In this case $\partial^2 P/\partial R^2 = -1/(8R_0^3) < 0$, and P is at a maximum. Its maximum value is

$$P = \frac{|E|^2}{4R_0}.$$

The conditions of the maximum dissipated power given by Eqs. (9.76) and (9.77) are called **impedance matching**.

◇

Exercises

9.1 Expand the following rectangular alternating current, one period of which is given by (see Fig. 9.29)

$$I(t) = I_0; \quad 0 \le t < \frac{T}{2},$$
$$= -I_0; \quad \frac{T}{2} \le t < T,$$

in a Fourier series with $\omega = 2\pi/T$.

9.2 Determine the impedance, admittance, resistance, reactance, conductance, and susceptance of the connected circuit elements shown in Fig. 9.30.

Fig. 9.29 Rectangular alternating current

Fig. 9.30 Connected circuit elements

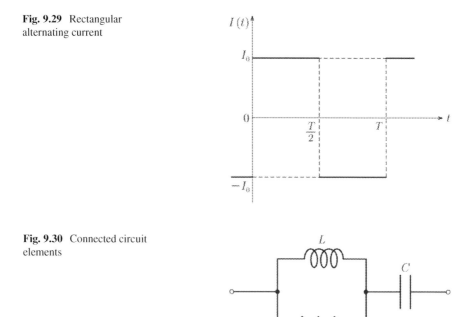

Fig. 9.31 Connected circuit elements

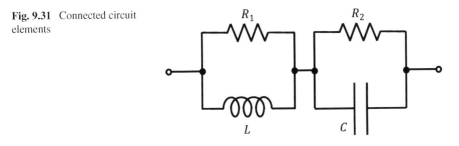

9.3 Represent resistance R and reactance X using conductance G and susceptance B.

9.4 Derive the condition so that the impedance of the connected circuit elements shown in Fig. 9.31 is a constant resistance independent of the angular frequency ω.

9.5 Determine R_4 and L_4 so that the voltage between B and C is zero in the AC bridge shown in Fig. 9.32.

9.6 Determine the angular frequency at which the impedance of the connected circuit elements in Fig. 9.33 is a resistance.

9.7 Determine the value of C of a variable capacitor so that the phase of the voltage V is ahead of that of E by $\pi/4$ in the circuit shown in Fig. 9.34.

Fig. 9.32 AC bridge

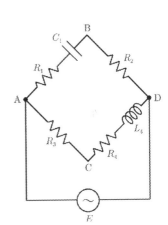

Fig. 9.33 Connected circuit elements

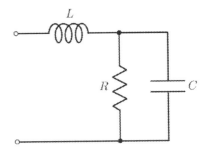

Fig. 9.34 AC circuit

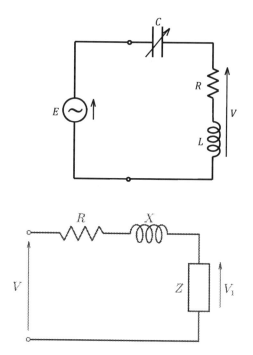

Fig. 9.35 AC circuit

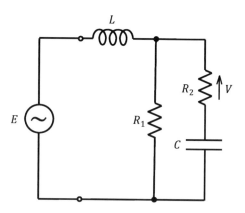

9.8 Suppose that the impedance of the load is $Z = |Z|e^{i\varphi}$ and $|Z|$ is a constant value in the circuit shown in Fig. 9.35. Determine the phase φ that minimizes the magnitude of the voltage $|V_1|$ of the load.

9.9 Determine the angular frequency at which the voltage $|V|$ of the resistor with R_2 takes on its maximum value in the circuit shown in Fig. 9.36.

Fig. 9.36 AC circuit

9.10 Determine the angular frequency characteristics of the phase of V_2/V_1 for the low-pass filter in Fig. 9.20 and for the high-pass filter in Fig. 9.22.

9.11 The circuit shown in Fig. 9.37 cuts off only the signal of the angular frequency of a designated area, and is called a **notch filter**. Determine the angular frequency characteristics of $|V_2/V_1|$ under the following given conditions:

$$L_1 = \frac{\Delta \omega R}{\omega_c^2}, \quad L_2 = \frac{R}{\Delta \omega}, \quad C_2 = \frac{1}{\Delta \omega R}, \quad C_2 = \frac{\Delta \omega}{\omega_c^2 R}.$$

Fig. 9.37 Notch filter circuit

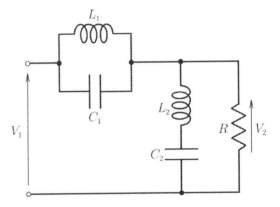

9.12 Determine the value of the resistance R at which the dissipated power takes on its maximum value in the circuit shown in Fig. 9.38.

Fig. 9.38 AC circuit

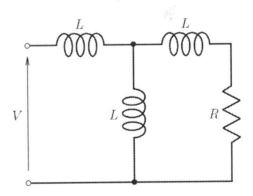

9.13 We are proving that the power in impedance Z in the circuit shown in Fig. 9.39 is given by

$$P = \frac{1}{2}R\left(|I_1|^2 - |I_2|^2 - |I_3|^2\right).$$

Answer the following questions:

(a) Represent $|I_1|^2$ and $|I_2|^2$ with $|I_3|^2$.
(b) Represent P with Z and I_3.
(c) Prove the proposition.

Fig. 9.39 AC circuit

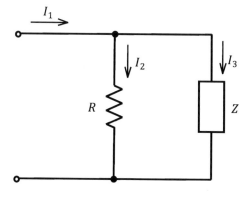

Chapter 10
Transformer Circuit

Abstract In this chapter we learn about the characteristics of the transformer, which is one of circuit components. Transformers are classified into voltage transformers that change voltage and current transformers that change current, depending on the magnitude of the impedance of the load connected to the transformer. The function of a transformer that involves transforming the load impedance is also introduced. It is possible to represent a circuit with a transformer as an equivalent circuit without a transformer. We treat various phenomena of AC circuits with transformers.

10.1 Transformer

A device used to change magnitudes of AC voltage or current is the **transformer**, the function of which was learned in Sect. 6.1. Suppose a transformer composed of two coils that are coupled magnetically, with the self-inductances of the primary and secondary coils denoted by L_1 and L_2, and the mutual inductance between them denoted by M, as shown in Fig. 10.1. When the voltages on these coils are V_1 and V_2 and the currents are I_1 and I_2, the relations between these quantities are described by

$$V_1 = i\omega L_1 I_1 + i\omega M I_2,$$
$$V_2 = i\omega M I_1 + i\omega L_2 I_2, \tag{10.1}$$

where resistance is disregarded. These equations are rewritten as

$$V_1 = i\omega(L_1 - M)I_1 + i\omega M(I_1 + I_2),$$
$$V_2 = i\omega M(I_1 + I_2) + i\omega(L_2 - M)I_2. \tag{10.2}$$

These conditions can be realized by a circuit without the transformer, as shown in Fig. 10.2. So, the circuits shown in Figs. 10.1 and 10.2 are equivalent to each other.

T. Matsushita, *Electricity*,
https://doi.org/10.1007/978-3-031-44002-1_10

Fig. 10.1 Transformer

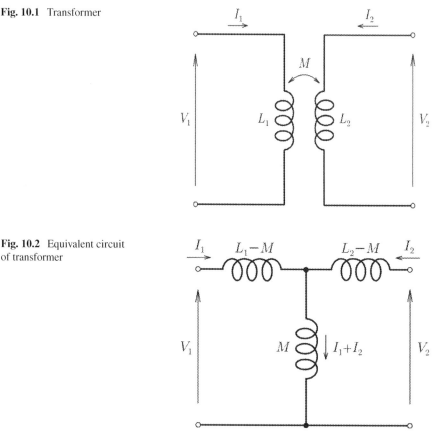

Fig. 10.2 Equivalent circuit
of transformer

As discussed in Chap. 6, the self-inductances L_1 and L_2 are defined to be positive, i.e., when the directions of the currents are defined, the directions of the magnetic flux are designated. On the other hand, the mutual inductance takes on a positive or negative value. There is a relation among these inductances given by

$$M^2 \leq L_1 L_2. \tag{10.3}$$

If we write this as

$$M = k(L_1 L_2)^{1/2}, \tag{10.4}$$

$|k|$ is called a **coupling coefficient**, which satisfies $|k| \leq 1$. The coupling in the case of $|k| = 1$ is called **tight coupling**.

Suppose that a load of impedance Z_2 is connected to the secondary winding, as shown in Fig. 10.3. In this case, the condition,

Fig. 10.3 Load connected to the secondary winging of a transformer

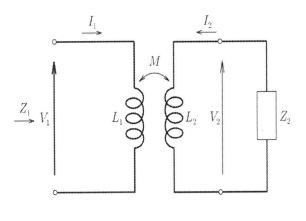

$$V_2 = -Z_2 I_2 \tag{10.5}$$

is added. Then, elimination of V_2 from the second equation of Eq. (10.1) yields

$$I_2 = -\frac{i\omega M}{i\omega L_2 + Z_2} I_1. \tag{10.6}$$

The first equation is also rewritten as

$$V_1 = \left(i\omega L_1 + \frac{\omega^2 M^2}{i\omega L_2 + Z_2} \right) I_1. \tag{10.7}$$

The impedance viewed from the primary coil, i.e., the input impedance, is given by

$$Z_1 = \frac{V_1}{I_1} = i\omega L_1 + \frac{\omega^2 M^2}{i\omega L_2 + Z_2} \equiv i\omega L_1 + Z, \tag{10.8}$$

and the corresponding equivalent circuit is shown in Fig. 10.4. So, the impedance is changed from Z_2 to Z_1 by inserting a transformer. Thus, the transformer has another function, to transform impedance.

Suppose that $|Z_2|$ is very large. Then, we have $V_1 = i\omega L_1 I_1$ from Eq. (10.7) and $V_2 = i\omega M I_1$ from Eqs. (10.5) and (10.6). Thus, we obtain

$$V_2 = \frac{M}{L_1} V_1. \tag{10.9}$$

Such a transformer used to change the voltage on a load is called a **voltage transformer**. On the other hand, if $|Z_2|$ is much smaller than ωL_2, Eq. (10.6) leads to

$$I_2 = -\frac{M}{L_2} I_1. \tag{10.10}$$

Fig. 10.4 Equivalent circuit
of Fig. 10.3. Z is defined in
Eq. (10.8)

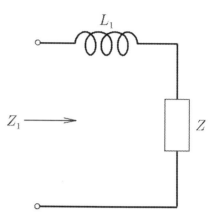

Such a transformer used to change the current flowing through a load is called a
current transformer.

Here, we estimate the energy stored in the transformer in Fig. 10.1 when the
currents I_1 and I_2 are applied to the primary and secondary circuits, respectively.
To make current I_1' flow through the primary circuit against the electromotive force
opposing it, it is necessary to apply an additional voltage $V_1' = L_1 dI_1'/dt$. The power
supplied by the power source is $V_1' I_1'$. Hence, the stored energy in the primary circuit
when current I_1 is applied is

$$U_{m1} = \int V_1' I_1' dt = L_1 \int_0^{I_1} I_1' dI_1' = \frac{1}{2} L_1 I_1^2. \tag{10.11}$$

Then, we apply current I_2 to the secondary circuit. The energy supplied by the power
source in the secondary circuit is similarly given by

$$U_{m2} = \frac{1}{2} L_2 I_2^2. \tag{10.12}$$

In the meantime, the electromotive force, $-MdI_2'/dt$, appears in the primary circuit,
which works to change the primary current. So, the power source in the primary
circuit must supply additional energy by

$$U_{m3} = \int I_1 M \frac{dI_2'}{dt} dt = M I_1 \int_0^{I_2} dI_2' = M I_1 I_2 \tag{10.13}$$

to keep the current I_1 flowing in the primary circuit. In the meantime, since the current
in the primary circuit does not change, the electromotive force does not appear in
the secondary circuit. As a result, the total stored energy is given by

$$U_m = U_{m1} + U_{m2} + U_{m3} = \frac{1}{2}\left(L_1 I_1^2 + 2M I_1 I_2 + L_2 I_2^2\right). \qquad (10.14)$$

In the general case of variation with time, Eq. (10.1) is given by

$$V_1' = L_1 \frac{dI_1'}{dt} + M \frac{dI_2'}{dt},$$
$$V_2' = M \frac{dI_1'}{dt} + L_2 \frac{dI_2'}{dt}, \qquad (10.15)$$

and the input power from the power sources of the two circuits is given by

$$P = V_1' I_1' + V_2' I_2' = \frac{1}{2} \cdot \frac{d}{dt}\left(L_1 I_1'^2 + 2M I_1' I_2' + L_2 I_2'^2\right). \qquad (10.16)$$

So, Eq. (10.14) can be directly obtained by integrating this with respect to time.

Example 10.1 Determine the impedance Z of the circuit shown in Fig. 10.5.

Fig. 10.5 Circuit with a transformer

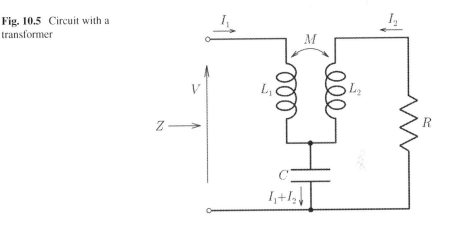

Solution 10.1 Since voltage V is applied to L_1 and C connected in series, we have

$$V = i\omega L_1 I_1 + i\omega M I_2 + \frac{1}{i\omega C}(I_1 + I_2)$$
$$= i\omega(L_1 - M)I_1 + \left(i\omega M + \frac{1}{i\omega C}\right)(I_1 + I_2).$$

For the right closed circuit composed of L_2, C, and R, we have

Fig. 10.6 Equivalent circuit
of the circuit shown in
Fig. 10.5

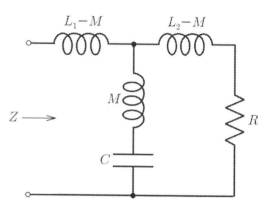

$$0 = i\omega L_2 I_2 + i\omega M I_1 + \frac{1}{i\omega C}(I_1 + I_2) + R I_2$$

$$= \left(i\omega M + \frac{1}{i\omega C}\right)(I_1 + I_2) + [R + i\omega(L_2 - M)]I_2.$$

Hence, the equivalent circuit of this circuit is shown in Fig. 10.6, and the impedance
is given by

$$Z = i\omega(L_1 - M) + \frac{[R + i\omega(L_2 - M)][i\omega M + 1/(i\omega C)]}{i\omega L_2 + R + 1/(i\omega C)}$$

$$= \frac{R(1 - \omega^2 L_2 C) + i\omega[(L_1 + L_2 - 2M) - \omega^2(L_1 L_2 - M^2)C]}{1 - \omega^2 L_2 C + i\omega C R}.$$

◇

10.2   Transformer Circuit

Here, we determine currents I_1 and I_2 in the transformer circuit shown in Fig. 10.7.
For the voltage on the left branch with inductance L_1, we have

$$E = i\omega L_1 I_1 + i\omega M I_2. \tag{10.17}$$

The relation on the right branch with inductance L_2 and resistance R is written as

$$E = i\omega L_2 I_2 + i\omega M I_1 + R I_2. \tag{10.18}$$

Then, the currents are determined to be

Fig. 10.7 AC circuit with a transformer

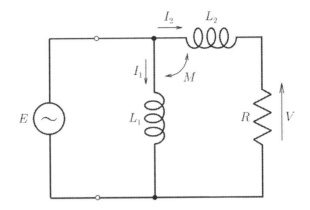

$$I_1 = -\frac{[R + i\omega(L_2 - M)]E}{\omega^2(L_1L_2 - M^2) - i\omega L_1 R},$$

$$I_2 = \frac{(L_1 - M)E}{L_1 R + i\omega(L_1L_2 - M^2)}. \tag{10.19}$$

We determine the condition that the voltage on the resistor R is in the same phase as the output voltage of the source. The voltage on the resistor R is given by

$$V = RI_2 = \frac{(L_1 - M)RE}{L_1 R + i\omega(L_1L_2 - M^2)}. \tag{10.20}$$

To realize that V/E is a positive real number, the conditions of the tight coupling,

$$L_1L_2 = M^2, \tag{10.21}$$

and

$$L_1 > M \tag{10.22}$$

must be satisfied.

Example 10.2 Show the equivalent circuit of the circuit shown in Fig. 10.7.

Solution 10.2 Equations (10.17) and (10.18) are rewritten, respectively, as

$$E = i\omega M(I_1 + I_2) + i\omega(L_1 - M)I_1,$$
$$E = i\omega M(I_1 + I_2) + i\omega(L_2 - M)I_2 + RI_2.$$

The equivalent circuit in which the above conditions are realized is shown in Fig. 10.8.

Fig. 10.8 Circuit equivalent
to the circuit shown in
Fig. 10.7

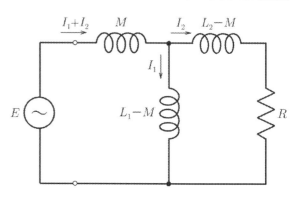

$$\Diamond$$

Example 10.3 Derive the condition in which the voltage between A and B is zero
in the circuit shown in Fig. 10.9.

Fig. 10.9 AC circuit with a
transformer

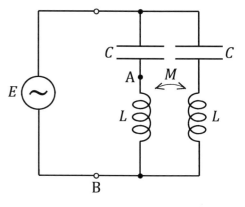

Solution 10.3 The circuit equivalent to the given circuit is shown in Fig. 10.10. If
we denote the whole current by I, the current in each branch is $I/2$. Thus, the voltage
between A and B is

$$V = \frac{i\omega(L - M)}{2}I + i\omega M I = \frac{i\omega(L + M)}{2}I.$$

The condition in which this voltage is zero is given by

$$L + M = 0.$$

Fig. 10.10 Circuit equivalent to the circuit in Fig. 10.9

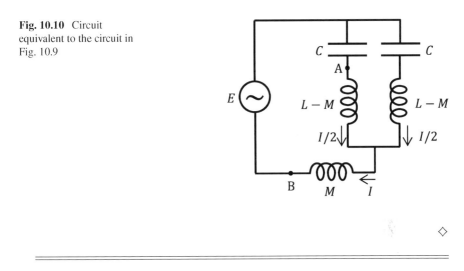

Example 10.4 Determine the condition in which the phase of current I_2 is ahead of that of I_1 by $\pi/2$ in the circuit shown in Fig. 10.11. Is it possible to attain the condition by changing the angular frequency ω?

Fig. 10.11 AC circuit with a transformer

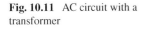

Solution 10.4 The circuit equivalent to this circuit is shown in Fig. 10.12. Since the voltage between A and B is common, we have

$$\left[i\omega(L_1 - M) + \frac{1}{i\omega C}\right]I_1 = [i\omega(L_2 - M) + R]I_2.$$

This leads to

$$\frac{I_2}{I_1} = \frac{i[\omega^2(L_1 - M) - C^{-1}]}{\omega R + i\omega^2(L_2 - M)}.$$

So that this is a positive imaginary number, the following conditions must be satisfied:

$$M = L_2,$$

$$L_1 - M - \frac{1}{\omega^2 C} > 0.$$

Thus, from the coupling condition, we have $L_1 > M$, and the proposition is realized for

$$\omega > \frac{1}{(L_1 - M)^{1/2} C^{1/2}}.$$

Fig. 10.12 Circuit equivalent to the circuit in Fig. 10.11

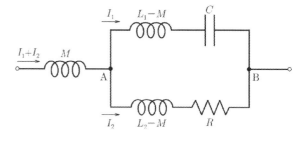

Exercises

10.1 Prove Eq. (10.3).

10.2 Show the circuit equivalent to that shown in Fig. 10.13, and determine its impedance.

10.3 Derive the condition in which the current flowing through the resistance R becomes 0 in the circuit shown in Fig. 10.5.

10.4 Determine the condition for R_4 and L_4 in which the voltage between A and B is zero in the circuit shown in Fig. 10.14. This is called the **Heaviside bridge**.

10.5 Determine the condition for M and ω in which $|I_1| = |I_2|$ and the phase difference between I_1 and I_2 is $\pi/2$ in the circuit shown in Fig. 10.15.

Fig. 10.13 AC circuit with a transformer

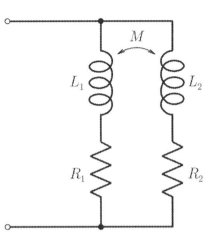

Fig. 10.14 Heaviside bridge

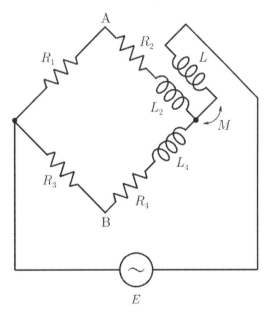

Fig. 10.15 AC circuit with a transformer

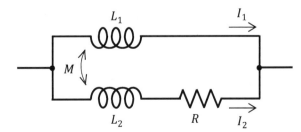

Chapter 11
Theorems for Electric Circuits

Abstract In this chapter we learn about various theorems for electric circuits for a greater understanding of the solution methods we have used. First, we learn Kirchhoff's law for the DC resistor circuit, and the branch current method, the closed current method, and the node potential method, which are used to determine flowing currents, are introduced. These methods can be extended to AC circuits. The principle of superposition, Thévenin's theorem, and Norton's theorem are also introduced, and the uses of these theorems are shown for various examples.

11.1 Kirchhoff's Law

Many problems have been solved for various DC and AC circuits. In this chapter, such solution methods are reviewed using fundamental analytic methods, and some theorems are introduced. These provide the basis for the solution methods that we have already used, and are useful to solve complicated problems.

Fundamental features of electric circuits are described by **Kirchhoff's law**. Here, we define the terms used in the law, before we go into details about the law. There are points at which currents pass through an electric circuit composed of power sources and circuit elements, and such points are called **nodes**. There is a passage for currents between adjacent two nodes, and these passages are called **branches**. An example is shown in Fig. 11.1. We can define a path which starts at a node, passes through each node and branch once, and goes back to the starting node. Such a path is called a **closed path**. The current that flows through each branch is a **branch current**, and the electric potential difference between two adjacent nodes is a **branch voltage**.

Kirchhoff's law consists of two laws, and the first law states as follows:

(a) The algebraic sum of currents entering or exiting an arbitrary node is zero. Here, a current that passes out is counted as positive. That is, if we denote the current passing out of the node through the i-th branch by $I_i (i = 1, 2, \ldots, n)$, as illustrated in Fig. 11.2, we have

$$\sum_{i=0}^{n} I_i = 0. \tag{11.1}$$

Fig. 11.1 Part of an electric
circuit

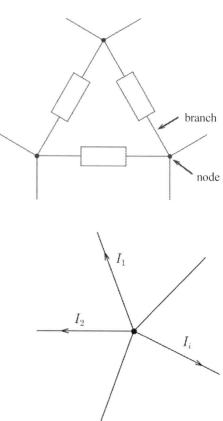

Fig. 11.2 Node and branch
currents

This holds also for AC currents, and is derived by applying Eq. (3.5) to a closed
surface that includes a corresponding node. The second law is stated for DC electric
circuits as follows:

(b) The algebraic sum of branch voltages is zero in an arbitrary closed path
     composed of branches in an electrical network. Hence, the sum of electromotive
     forces is equal to the sum of voltage drops. If we denote by $E_i (i = 1, 2, \ldots, m)$
     and $V_j (j = 1, 2, \ldots, n)$ the electromotive forces and the voltage drops in the
     closed path, as shown in Fig. 11.3, we have

$$\sum_{i=1}^{m} E_i = \sum_{j=1}^{n} V_j. \tag{11.2}$$

This can be derived by application of Eq. (3.18) to the indicated closed path. In
power sources the direction of the electric field is opposite to that of the current. The
extension of the second law to AC circuits will be mentioned later.

Fig. 11.3 Closed path in
electric circuit

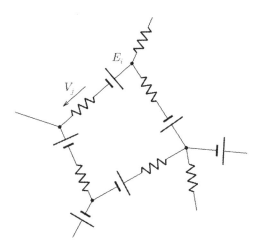

11.2 Circuit Equation

In the former sections we used the currents flowing in each branch. These branch
currents can be expressed by superimposing currents flowing in closed paths. Suppose
that an electric circuit is composed of 4 nodes and 6 branches, as shown in Fig. 11.4a.
There are six branch currents, but these are expressed by superposition of three closed
currents, I_1, I_2, and I_3. Namely,

$$I_{12} = I_1, \quad I_{13} = -I_3, \quad I_{14} = -I_1 + I_3,$$
$$I_{23} = I_2, \quad I_{24} = I_1 - I_2, \quad I_{34} = I_2 - I_3. \tag{11.3}$$

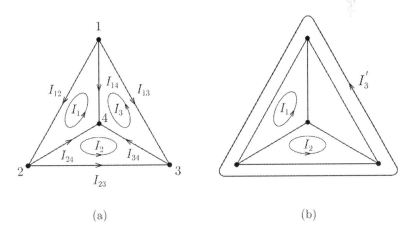

(a)   (b)

Fig. 11.4 **a** Example of branch currents and closed currents and **b** another choice of closed currents

This results from the fact that all the branch currents are not independent of the others due to the relation given by Eq. (11.1). For example, when I_{12} and I_{14} are given, I_{13} is determined, and if I_{34} is given, the remaining I_{23} and I_{24} are determined. When the numbers of nodes and branches are n and b, respectively, there are n node equations. Since when all the node equations are summed, the sum is reduces to zero. So, the number of independent node equations is $n - 1$. The number of independent branch currents is reduced by this number of independent node equations and is given by $b - n + 1$. We have $n = 4$ and $b = 6$ for the circuit shown in Fig. 11.4a, and the number of independent branch currents is 3. This is the reason why we can express all the branch currents with a smaller number of closed currents. So, it is in general easier to use closed currents as unknown variables. The choice of independent closed currents is not unique and we can use I_3' instead of I_3, as shown in Fig. 11.4b. It is required that all the branches must be passed by some currents and that all the closed currents must be independent of the others.

From the above points, we have the **branch current method** and the **closed current method** for describing circuit equations. Another method is the **node potential method** in which the node potentials are used as independent variables. In the following sections, these methods are introduced.

11.2.1 Branch Current Method

Here, we determine the currents using the branch current method for the circuit shown in Fig. 11.5. Since the number of nodes is 2, the number of independent Eqs. (11.1) is 1. So, if we denote the branch currents as shown in the figure, the equation on node A is given by

Fig. 11.5 DC circuit

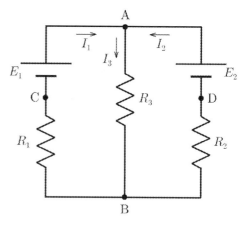

$$-I_1 - I_2 + I_3 = 0. \tag{11.4}$$

The equation on node B is the same one with different signs. In the case of the closed paths for Eq. (11.2), the number of independent closed paths is 2, and we use ABC and ABD. Equation (11.2) for ABC is

$$E_1 = R_1 I_1 + R_3 I_3, \tag{11.5}$$

and that for ABD is

$$E_2 = R_2 I_2 + R_3 I_3. \tag{11.6}$$

Eliminating I_3 by using Eq. (11.4), Eqs. (11.5) and (11.6) lead, respectively, to

$$E_1 = (R_1 + R_3)I_1 + R_3 I_2, \tag{11.7}$$

$$E_2 = R_3 I_1 + (R_2 + R_3)I_2. \tag{11.8}$$

Thus, all the currents are determined to be

$$
\begin{aligned}
I_1 &= \frac{(R_2 + R_3)E_1 - R_3 E_2}{R_1 R_2 + R_2 R_3 + R_3 R_1}, \\
I_2 &= \frac{-R_3 E_1 + (R_1 + R_3)E_2}{R_1 R_2 + R_2 R_3 + R_3 R_1}, \\
I_3 &= \frac{R_2 E_1 + R_1 E_2}{R_1 R_2 + R_2 R_3 + R_3 R_1}.
\end{aligned} \tag{11.9}
$$

The electric circuit in Fig. 7.22 contains 4 nodes and 6 branches. Hence, the numbers of independent currents and closed paths are 3, and the problem can be solved using three currents, I_1, I_3, and I. Equations (7.25), (7.26), and (7.17) describe the second law for a closed path including the power source and ABC, including the power source and ACD, and including ABC, respectively.

Example 11.1 Determine the currents using the branch current method for the circuit shown in Fig. 11.6.

Fig. 11.6 DC circuit

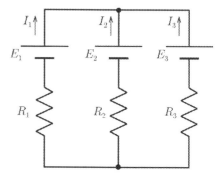

Solution 11.1 If we denote the currents as in the figure, we have $I_3 = -I_1 - I_2$ and

$$E_1 - E_2 = R_1 I_1 - R_2 I_2$$

for the closed path including the DC voltage sources E_1 and E_2, and

$$E_3 - E_2 = -R_2 I_2 + R_3 I_3$$

for the closed path including the DC voltage sources E_2 and E_3. Substituting I_3 into these equations, each current is determined to be

$$I_1 = \frac{(R_2 + R_3)E_1 - R_3 E_2 - R_2 E_3}{R_1 R_2 + R_2 R_3 + R_3 R_1},$$
$$I_2 = \frac{-R_3 E_1 + (R_1 + R_3)E_2 - R_1 E_3}{R_1 R_2 + R_2 R_3 + R_3 R_1},$$
$$I_3 = \frac{R_2 E_1 + R_1 E_2 + (R_1 + R_2)E_3}{R_1 R_2 + R_2 R_3 + R_3 R_1}.$$

$\diamondsuit$

Next, we treat the case in which a current source is included. Suppose the circuit shown in Fig. 11.7. When the branch currents are defined as shown in the figure, we have

$$I_3 = I_1 + I_2. \tag{11.10}$$

Application of the second law to the closed path composed of R_1 and R_3 yields

$$E_1 = R_1 I_1 + R_3 I_3. \tag{11.11}$$

The current source is inevitably included in other closed paths. It should be noted that the second law cannot be applied for this path, since the electromotive force of

Fig. 11.7 DC circuit with
current source

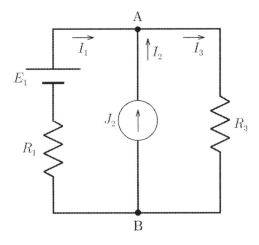

the current source is unknown. Instead of it, current I_2 is automatically given by

$$I_2 = J_2. \tag{11.12}$$

Substitution of this into Eqs. (11.10) and (11.11) yields

$$I_1 = \frac{E_1 - R_3 J_2}{R_1 + R_3},$$
$$I_3 = \frac{E_1 + R_1 J_2}{R_1 + R_3}. \tag{11.13}$$

Thus, in the case where the current source is included, the current is directly given, while Eq. (11.2) cannot be used, since the electromotive force of the current source changes depending on surrounding resistances. Kirchhoff's second law holds as a result, however.

11.2.2 Closed Current Method

Here, we introduce the solution method with closed currents. Since the paths of closed currents are closed, the conservation law of current, Eq. (11.1), holds automatically, even when superimposing them. We denote the closed currents as shown in Fig. 11.8 for the circuit shown in Fig. 11.5. Then, along the path for closed current I_A, the second law is described as

$$E_1 = R_1 I_A + R_3(I_A + I_B). \tag{11.14}$$

Similarly, we have

Fig. 11.8 Solution with
closed current method

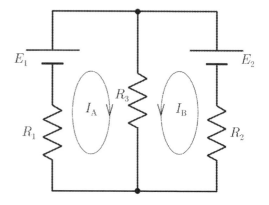

$$E_2 = R_2 I_B + R_3(I_A + I_B) \tag{11.15}$$

for the closed path of I_B. If we note that

$$I_1 = I_A, I_2 = I_B, I_3 = I_A + I_B, \tag{11.16}$$

we find that Eqs. (11.14) and (11.15) are the same as Eqs. (11.5) and (11.6), respectively. Hence, the solution, Eq. (11.9), is obtained.

Example 11.2 Determine the closed currents, I_1, I_2, and I_3, in the circuit shown in Fig. 11.9.

Fig. 11.9 DC circuit

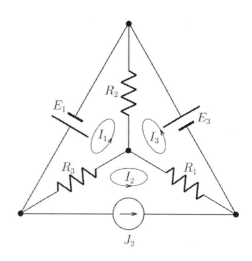

Solution 11.2 Closed current I_2 is simply given by

$$I_2 = J_2.$$

The second law for the paths of I_1 and I_3 are, respectively, given by

$$E_1 = R_2(I_1 - I_3) + R_3(I_1 - I_2),$$
$$E_3 = R_1(I_3 - I_2) + R_2(I_3 - I_1).$$

Eliminating I_3 using the above equation with substitution of I_2 by J_2, we have

$$I_1 = J_2 + \frac{(R_1 + R_2)E_1 + R_2 E_3}{R_1 R_2 + R_2 R_3 + R_3 R_1}.$$

Then, I_3 is determined to be

$$I_3 = J_2 + \frac{R_2 E_1 + (R_2 + R_3)E_3}{R_1 R_2 + R_2 R_3 + R_3 R_1}.$$

$\diamond$

11.2.3 Node Potential Method

It is also possible to solve problems by using electric potentials at each node instead of branch currents or closed currents. In this case, Eq. (1.17) holds for a closed loop, and it is not necessary to use the second law. Each branch current is determined by the electric potentials on both sides, the electromotive force of the voltage source, and the resistance in the corresponding branch.

Suppose the circuit shown in Fig. 11.5. We denote the electric potential at node A by V, setting the electric potential at node B to be 0. Then, currents I_1, I_2, and I_3 are given by $(E_1 - V)/R_1$, $(E_2 - V)/R_2$, and V/R_3, respectively. Hence, Eq. (11.4) leads to

$$-\frac{E_1 - V}{R_1} - \frac{E_2 - V}{R_2} + \frac{V}{R_3} = 0. \tag{11.17}$$

Thus, we have

$$V = \frac{R_2 R_3 E_1 + R_3 R_1 E_2}{R_1 R_2 + R_2 R_3 + R_3 R_1}. \tag{11.18}$$

Then, the currents in Eq. (11.9) are determined. The equation describing the relation on currents, such as Eq. (11.17), is called a nodal equation.

Fig. 11.10 DC circuit

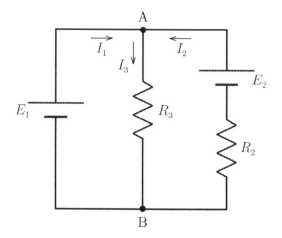

In the case where the number of nodes is n, there are $n - 1$ independent node potentials. So, these unknown variables can be solved by using $n - 1$ nodal equations. If the power source is a current source, the branch current is given by the current of the current source, independently of the potential difference between the two nodes.

If there is a voltage source only in a branch, as shown in Fig. 11.10, the potential difference between the two nodes is determined by the electromotive force of the voltage source, but the branch current is not simply determined. This is because there is no voltage drop due to the branch current. That is, even if we describe the second law for a closed path including the voltage source with E_1, branch current I_1 does not appear in it. The value of this current is determined by the values of the resistances around it. Here, we show the concrete solution method. You should solve this problem with both the branch current and the closed current methods (see Exercises 11.4 and 11.5). We set the electric potential at node B to zero. Then, the electric potential at node A is E_1. Then, currents I_2 and I_3 are determined to be $I_2 = (E_2 - E_1)/R_2$ and $I_3 = E_1/R_3$, respectively. So, the current I_1 is determined to be

$$I_1 = I_3 - I_2 = \frac{(R_2 + R_3)E_1 - R_3 E_2}{R_2 R_3}. \qquad (11.19)$$

Example 11.3 Determine the branch currents in the circuit shown in Fig. 11.7 using the node potential method.

Solution 11.3 We set the electric potential at node B to be zero and denote by V the electric potential at node A. Then, the currents, I_1, I_2, and I_3, are given by $(E_1 - V)/R_1$, J_2, and V/R_3, respectively. Then, Eq. (11.4) leads to

$$-\frac{E_1 - V}{R_1} - J_2 + \frac{V}{R_3} = 0,$$

and V is determined to be

$$V = \frac{R_3(E_1 + R_1 J_2)}{R_1 + R_3}.$$

Then, each current is obtained as in Eq. (11.13).

◇

Example 11.4 We denote by V_A and V_B the electric potentials at nodes A and B, respectively, in the circuit shown in Fig. 11.11. Determine current I by using the node potential method.

Fig. 11.11 DC circuit

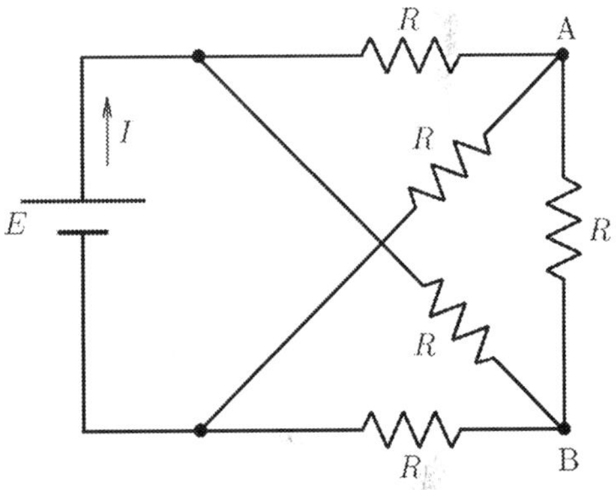

Solution 11.4 From the fact that the total current exiting point A is zero, we have

$$\frac{V_A - E}{R} + \frac{V_A}{R} + \frac{V_A - V_B}{R} = 0,$$

which is reduced to

$$3V_A - V_B = E.$$

From the fact that the total current that exists point B is zero, we have similarly

$$\frac{V_B - E}{R} + \frac{V_B - V_A}{R} + \frac{V_B}{R} = 0,$$

which is reduced to

$$-V_A + 3V_B = E.$$

Then, we have $V_A = V_B = E/2$. That is, nodes A and B are equipotential. Thus, the total current is determined to be

$$I = \frac{E - V_A}{R} + \frac{E - V_B}{R} = \frac{E}{R}.$$

The equipotential between nodes A and B can be intuitively derived, as shown in Chap. 7.

◇

11.3   Extension to AC Circuit

We have learned the characteristics of AC circuits in Chaps. 8–10. Here, we properly extend Kirchhoff's law used in DC circuits to AC circuits.

Since the conservation law for currents holds also for AC, the first law, Eq. (11.1), still holds. On the other hand, the electrostatic potential cannot be used because of electromagnetic induction. That is, since Eq. (1.17) is not satisfied, the second law does not hold. For example, in the circuit shown in Fig. 11.12, the electrostatic field integrated along the left closed path leads to

$$-E(t) + R_1 I_1(t) + \frac{1}{C_1} \int I_1(t) dt. \tag{11.20}$$

Fig. 11.12   AC circuit

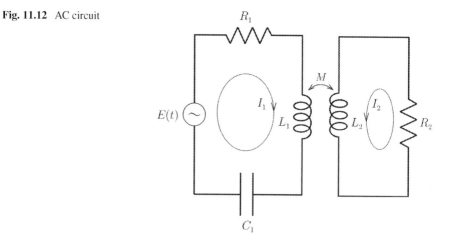

This is not zero because of the electromagnetic induction, but is given by

$$-L_1 \frac{dI_1(t)}{dt} - M \frac{dI_2(t)}{dt}. \tag{11.21}$$

If we describe the above equation in a similar form as the second law, we have

$$E(t) = R_1 I_1(t) + \frac{1}{C_1} \int I_1(t)dt + L_1 \frac{dI_1(t)}{dt} + M \frac{dI_2(t)}{dt}. \tag{11.22}$$

From a theoretical viewpoint, since the third and fourth terms are electromotive forces, these terms should be moved to the left side. But these terms are regarded as potential drops caused by the power source to maintain current $I_1(t)$ against the electromagnetic induction. Thus, we obtain an equation of the same form as the second law by rewriting the induced electromotive forces as potential drops. In this case, the equation for the right closed path in Fig. 11.12 is similarly written as

$$0 = R_2 I_2(t) + L_2 \frac{dI_2(t)}{dt} + M \frac{dI_1(t)}{dt}. \tag{11.23}$$

These are the equations we have used. In the case of AC currents with angular frequency ω, these reduce to

$$E = \left(R_1 + i\omega L_1 + \frac{1}{i\omega C_1} \right) I_1 + i\omega M I_2, \tag{11.24}$$

$$0 = i\omega M I_1 + (R_2 + i\omega L_2) I_2. \tag{11.25}$$

Then, the currents are determined to be

$$I_1 = \frac{(R_2 + i\omega L_2)E}{R_1 R_2 - \omega^2 (L_1 L_2 - M^2) + L_2/C_1 + i[\omega(L_1 R_2 + L_2 R_1) - R_2/(\omega C_1)]}, \tag{11.26}$$

$$I_2 = -\frac{i\omega M E}{R_1 R_2 - \omega^2 (L_1 L_2 - M^2) + L_2/C_1 + i[\omega(L_1 R_2 + L_2 R_1) - R_2/(\omega C_1)]}. \tag{11.27}$$

From the above result, the second law is extended for a closed path that does not include current sources as

(c) If we denote by $E_i (i = 1, 2, \ldots, l)$, $V_j (j = 1, 2, \ldots, m)$, and $V'_k (k = 1, 2, \ldots, n)$ the electromotive forces, the voltage drops, and the voltage drops due to the electromagnetic induction in the given circuit, respectively, the extended second law is given by

$$\sum_{i=1}^{l} E_i = \sum_{j=1}^{m} V_j + \sum_{k=1}^{n} V_k'. \tag{11.28}$$

Example 11.5 Describe the second law for the AC circuit shown in Fig. 11.13.

Fig. 11.13 AC circuit

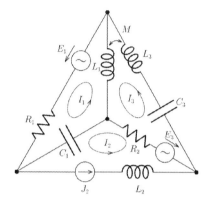

Solution 11.5 For the closed path for current I_1, we have

$$E_1 = R_1 I_1 + \frac{1}{i\omega C_1}(I_1 - I_2) + i\omega L_1(I_1 - I_3) + i\omega M I_3$$

$$= \left(R_1 + i\omega L_1 + \frac{1}{i\omega C_1} \right) I_1 - \frac{1}{i\omega C_1} I_2 - i\omega(L_1 - M)I_3.$$

For the closed path for current I_2, we have simply

$$J_2 = I_2.$$

The equation for the closed path for current I_3 is given by

$$E_3 = i\omega L_1(-I_1 + I_3) - i\omega M I_3 + R_2(-I_2 + I_3)$$

$$+ \frac{1}{i\omega C_3} I_3 + i\omega L_3 I_3 + i\omega M (I_1 - I_3)$$

$$= -i\omega(L_1 - M)I_1 - R_2 I_2 + \left[R_2 + i\omega(L_1 + L_3 - 2M) + \frac{1}{i\omega C_3} \right] I_3.$$

Confirm the above results using the equivalent circuit shown in Sect. 10.1.

◇

11.4 Principle of Superposition

The systems that we have treated are linear ones: In these systems, if the electromotive force of the voltage source is doubled, the current is also doubled. Hence, if there are some voltage sources in a circuit, the total current is given by the sum (superposition) of the current given by each voltage source. The same thing can be said for current sources. Thus, the following law holds [1].

Suppose that voltage sources, $E_1, E_2, \ldots, E_m$, and current sources, $J_{m+1}, J_{m+2}, \ldots, J_{m+n}$, are included in a circuit. The branch voltage V_i and branch current I_i in the i-th branch are respectively given by the sum of V_{ij} and I_{ij} with respect to j, where V_{ij} and I_{ij} are the values of the branch voltage and current when the effects of power sources except for the j-th source are eliminated. That is,

$$V_i = \sum_{j=1}^{m+n} V_{ij}, \quad I_i = \sum_{j=1}^{m+n} I_{ij}. \tag{11.29}$$

In the above, the elimination of the effect of power sources means assuming short circuit for voltage sources and removal of current sources, leaving an opening in their place, to make the electromotive force and the current zero. This law is called the **principle of superposition**.

Here, we show that this law holds for the circuit shown in Fig. 11.5. The circuit in which voltage source E_1 remains and voltage source E_2 is short-circuited is shown in Fig. 11.14a, and the circuit in which voltage source E_2 remains and voltage source E_1 is short-circuited is shown in Fig. 11.14b. Each branch current in Fig. 11.14a is easily solved as

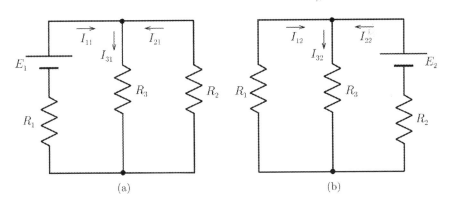

(a)   (b)

Fig. 11.14 **a** DC circuit in which voltage source E_2 is short-circuited and **b** DC circuit in which voltage source E_1 is short-circuited

$$I_{11} = \frac{(R_2 + R_3)E_1}{R_1 R_2 + R_2 R_3 + R_3 R_1},$$

$$I_{21} = -\frac{R_3 E_1}{R_1 R_2 + R_2 R_3 + R_3 R_1},$$

$$I_{31} = \frac{R_2 E_1}{R_1 R_2 + R_2 R_3 + R_3 R_1}. \tag{11.30}$$

Each branch current in Fig. 11.14b is also similarly obtained as

$$I_{12} = -\frac{R_3 E_2}{R_1 R_2 + R_2 R_3 + R_3 R_1},$$

$$I_{22} = \frac{(R_1 + R_3)E_2}{R_1 R_2 + R_2 R_3 + R_3 R_1},$$

$$I_{32} = \frac{R_1 E_2}{R_1 R_2 + R_2 R_3 + R_3 R_1}. \tag{11.31}$$

Using these results, we have

$$I_1 = I_{11} + I_{12} = \frac{(R_2 + R_3)E_1 - R_3 E_2}{R_1 R_2 + R_2 R_3 + R_3 R_1},$$

$$I_2 = I_{21} + I_{22} = \frac{-R_3 E_1 + (R_1 + R_3)E_2}{R_1 R_2 + R_2 R_3 + R_3 R_1},$$

$$I_3 = I_{31} + I_{32} = \frac{R_2 E_1 + R_1 E_2}{R_1 R_2 + R_2 R_3 + R_3 R_1}. \tag{11.32}$$

These results agree with those in Eq. (11.9). It is seen that the principle of superposition holds also for the branch voltage between A and B.

Example 11.6 Determine the branch currents for the circuit with a current source shown in Fig. 11.7 using the principle of superposition.

Solution 11.6 The circuit with the current source removed is shown in Fig. 11.15a, and that in which the voltage source is short-circuited is shown in Fig. 11.15b. Each branch current in Fig. 11.15a is obtained as

$$I_{11} = \frac{E_1}{R_1 + R_3}, \quad I_{21} = 0, \quad I_{31} = \frac{E_1}{R_1 + R_3},$$

and each branch current in Fig. 11.15b is similarly obtained as

$$I_{12} = -\frac{R_3 J_2}{R_1 + R_3}, \quad I_{22} = J_2, \quad I_{32} = \frac{R_1 J_2}{R_1 + R_3}.$$

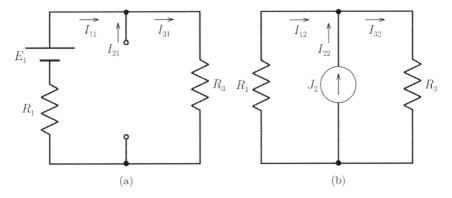

Fig. 11.15 **a** DC circuit with opened current source and **b** that with short-circuited voltage source

Thus, the branch currents are determined to be

$$I_1 = I_{11} + I_{12} = \frac{E_1 - R_3 J_2}{R_1 + R_3},$$

$$I_2 = I_{21} + I_{22} = J_2,$$

$$I_3 = I_{31} + I_{32} = \frac{E_1 + R_1 J_2}{R_1 + R_3}.$$

These agree with Eqs. (11.12) and (11.13).

$\diamondsuit$

11.5 Thévenin's Theorem

Suppose that electric circuit N_0 shown in Fig. 1.16a contains power sources inside it and that the voltage between the terminals a and b is V_0. When the effects of the

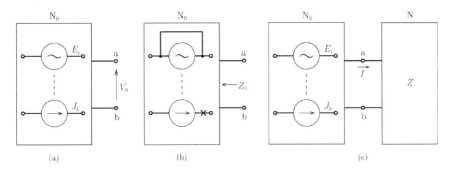

Fig. 11.16 **a** Circuit N_0 with power sources, **b** condition in which the effects of internal power sources are eliminated, and **c** condition in which circuit N of impedance Z without internal power sources is connected

internal power sources are eliminated by short-circuiting the voltage sources and by removal of the current sources with opening as shown in Fig. 11.16b, the impedance between the terminals a and b is assumed to be Z_0. When electric circuit N of impedance Z that does not contain power sources is connected to the terminals a and b as shown in Fig. 11.16c, the current that flows from the terminal a to the circuit N is given by

$$I = \frac{V_0}{Z_0 + Z}. \tag{11.33}$$

This is called **Thévenin's theorem** [1].

This theorem is proved here. The electric circuit shown in Fig. 11.16c is the same as the circuit with insertion of two voltage sources of electromotive force V_0 with opposite directions, as shown in Fig. 11.17. The principle of superposition proves that this circuit is equal to the superposition of the two electric circuits shown in Fig. 11.18a and b. All power sources in circuit N_0 are active in Fig. 11.18a, and all power sources in circuit N_0 are killed in Fig. 11.18b. Since the voltage between terminals a and b is zero in the circuit in Fig. 11.18a, the current I' that flows from terminal a into circuit N is zero. On the other hand, the left side of terminals a and b is just an impedance Z_0 in Fig. 11.18b. So, the current I'' that flows from terminal a into circuit N is given by $V_0/(Z_0+Z)$. Then, we have $I = I'+I''$ from superposition [1]. Thus, Eq. (11.33) is proved.

Thévenin's theorem shows that any circuit can be regarded as a power source with an electromotive force and an internal impedance, if we look inside from two arbitrary terminals. Figure 11.19a shows its equivalent circuit, and Fig. 11.19b is another expression with a current source.

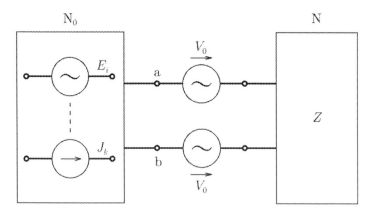

Fig. 11.17 Condition in which two voltage sources of electromotive force V_0 are inserted in the opposite directions in the electric circuit shown in Fig. 11.16c

Fig. 11.18 **a** Condition in which all power sources are active in circuit N_0 and one voltage source is connected and **b** condition in which all power sources are eliminated in circuit N_0 and another voltage source is connected

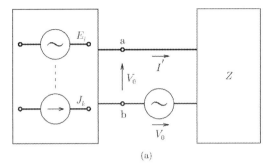

Fig. 11.19 Equivalent circuit of power source with **a** voltage source and **b** current source

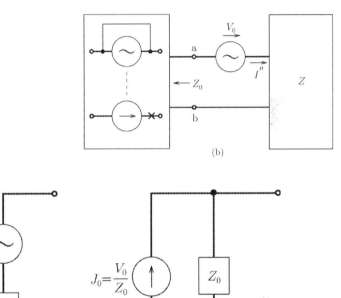

Here, we determine current I_3 flowing through R_3 in the circuit shown in Fig. 11.5 by using Thévenin's theorem. The circuit in which R_3 is removed is shown in Fig. 11.20a. Since the current flowing through this resistor is $I' = (E_1 - E_2)/(R_1 + R_2)$, the open-circuited voltage is given by

$$V_0 = E_1 - R_1 I' = \frac{R_2 E_1 + R_1 E_2}{R_1 + R_2}. \tag{11.34}$$

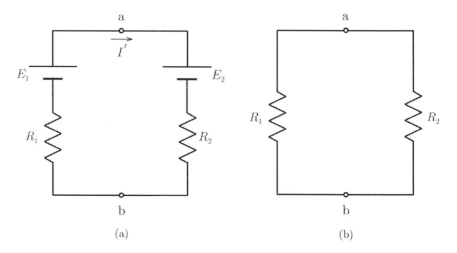

Fig. 11.20 a Circuit in which R_3 is removed from the circuit in Fig. 11.5 and **b** circuit with short-circuited voltage sources

On the other hand, the circuit in which the voltage sources are short-circuited in the circuit in Fig. 11.20a is shown in Fig. 11.20b. The resistance between terminals a and b is

$$R_0 = \frac{R_1 R_2}{R_1 + R_2}. \tag{11.35}$$

Then, from Eq. (11.33), current I_3 is determined to be

$$I_3 = \frac{V_0}{R_0 + R_3} = \frac{R_2 E_1 + R_1 E_2}{R_1 R_2 + R_2 R_3 + R_3 R_1}, \tag{11.36}$$

which agrees with Eq. (11.9).

Example 11.7 Determine current I in the circuit shown in Fig. 7.22 by using Thévenin's theorem.

Solution 11.7 The circuit in which a resistor (R) is removed is the same as the circuit shown in Fig. 7.14. The open-circuited voltage is given by Eq. (7.20). On the other hand, the circuit in which the voltage source is short-circuited is shown in Fig. 11.21. The resistance between terminals B and C is

$$R_0 = \frac{R_1 R_2}{R_1 + R_2} + \frac{R_3 R_4}{R_3 + R_4}.$$

So, using Thévenin's theorem, the current is determined to be

$$I = \frac{V_{BC}}{R_0 + R}$$

$$= \frac{(R_2 R_3 - R_4 R_1)E}{R_1 R_2 (R_3 + R_4) + R_3 R_4 (R_1 + R_2) + R(R_1 + R_2)(R_3 + R_4)}.$$

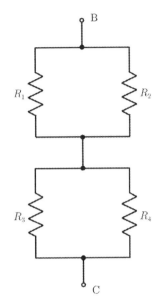

Fig. 11.21 Circuit in which the voltage source is short-circuited in the circuit shown in Fig. 7.14

◇

Norton's theorem forms a pair with Thévenin's theorem. It is stated as follows: Suppose a circuit N_0 with power sources, as shown in Fig. 11.22a. We denote by I_0 the current when the terminals a and b are short-circuited, and by Y_0 the admittance between these terminals when the effects of internal power sources are eliminated, as shown in Fig. 11.22b. When circuit N of admittance Y that does not contain power sources is connected to these terminals, as shown in Fig. 11.22c, the voltage between these terminals is given by

$$V = \frac{I_0}{Y_0 + Y}. \tag{11.37}$$

This theorem is now proved. The circuit shown in Fig. 11.22c is the same as the circuit with insertion of two current sources each generating current I_0 with opposite directions, as shown in Fig. 11.23. The principle of superposition proves that this circuit is equal to two superposed electric circuits, as shown in Fig. 11.24a and b. All power sources in circuit N_0 are active in Fig. 11.24a, and all power sources in

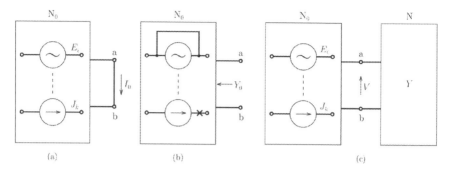

Fig. 11.22 a Circuit N_0 with power sources, **b** condition in which the effects of internal power sources are eliminated, and **c** condition in which circuit N of admittance Y without internal power sources is connected

circuit N_0 are eliminated in Fig. 11.24b. The voltage V' between terminals a and b is zero in Fig. 11.24a. On the other hand, the left side of terminals a and b is just an admittance Y_0 in Fig. 11.24b. So, the voltage V'' between terminals a and b is given by $I_0/(Y_0 + Y)$. Then, we have $V = V' + V''$ from superposition. Thus, Eq. (11.37) is proved.

The **Hoashi-Millman theorem** is an extension of Norton's theorem. It is stated as follows: The open-circuited voltage in an electric circuit composed of voltage sources and admittances, which is shown in Fig. 11.25, is given by

$$V_0 = \frac{\sum_{i=1}^{n} Y_i E_i}{\sum_{i=1}^{n} Y_i}. \tag{11.38}$$

This is proved by using Norton's theorem but it can be more simply proved with the node potential method. The current from each branch that flows through the terminal

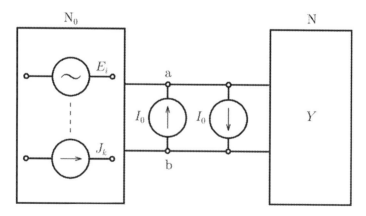

Fig. 11.23 Condition in which two current sources each generating current I_0 are inserted in the opposite directions in the electric circuit shown in Fig. 11.22c

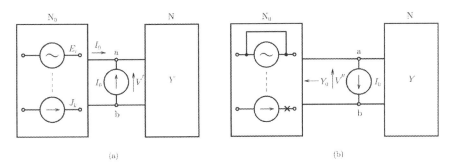

Fig. 11.24 a Condition in which all power sources are active in circuit N_0 and one current source is connected, and b condition in which all power sources are eliminated in circuit N_0 and another current source is connected

Fig. 11.25 Electric circuit composed of voltage sources and admittances

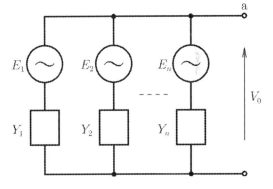

a is $Y_i(V_0 - E_i)$. Since the sum of these currents must be zero, we have

$$V_0 \sum_{i=1}^{n} Y_i = \sum_{i=1}^{n} Y_i E_i. \qquad (11.39)$$

Thus, Eq. (11.38) is proved.

Exercises

11.1 Describe circuit equations for the circuit shown in Fig. 11.26 using the branch current method.

11.2 Determine each current in the circuit shown in Fig. 11.27. Determine the condition in which $I_6 = 0$ is obtained when $E_1 = E_2 = E$ and $R_4 = R_5 = R$.

11.3 Describe the equations to determine each closed current in the resistor circuit shown in Fig. 7.22.

11.4 Determine branch currents I_1, I_2, and I_3 in the DC circuit shown in Fig. 11.10 using the branch current method.

Fig. 11.26 DC resistor
circuit

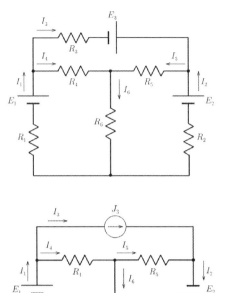

Fig. 11.27 DC resistor
circuit

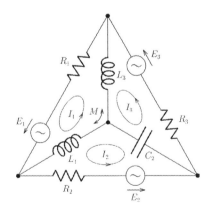

11.5 Determine current I_1 in the DC circuit shown in Fig. 11.10 using the closed
current method.

11.6 Determine current I in the DC circuit shown in Fig. 11.11 using the closed
current method.

11.7 Describe the circuit equations for the AC circuit shown in Fig. 11.28.

Fig. 11.28 AC circuit

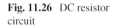

Fig. 11.29 Circuit

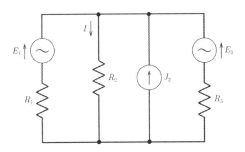

11.8 Determine current I in the DC circuit shown in Fig. 11.29 using the principle of superposition.

11.9 Figure 11.30 shows a resistor network of resistors with resistance R. Determine the resistance between terminals P and Q by calculating the voltage between the terminals when a DC current source with 1 A is connected between them, using the principle of superposition. (*Hint*: The given condition is obtained by superposing the condition in which the same current source is connected between terminal P and infinity and the condition in which the same current source is connected between infinity and terminal Q.)

11.10 Determine the current flowing through the resistor (R) in the circuit shown in Fig. 11.31 using Thévenin's theorem.

Fig. 11.30 Resistor network

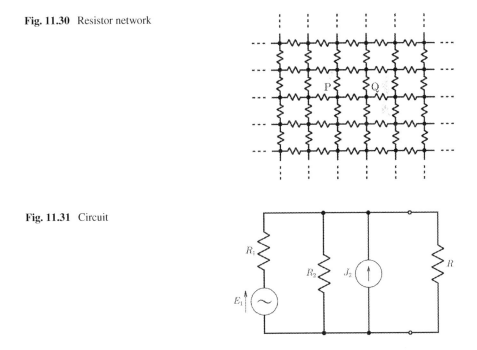

Fig. 11.31 Circuit

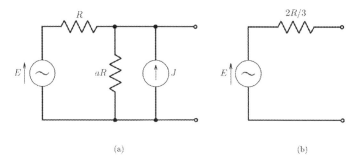

(a)   (b)

Fig. 11.32 Equivalent two circuits

11.11   When the equivalent circuit of the power source shown in Fig. 11.32a is the
circuit shown in (b), determine a and J.

11.12   Prove the Hoashi-Millman theorem using Norton's theorem.

Reference

1. Ohno K (1970) In: Sakakibara Y, Ohno K, Ozaki H (eds) Electric circuit I. Ohmsha (point:
various theorems) (in Japanese)

Appendix

A.1 Vector Analysis

A.1.1 Differentiation

When $f(x, y, z)$ is a given scalar function,

$$\text{grad } f = \frac{\partial f}{\partial x}\boldsymbol{i}_x + \frac{\partial f}{\partial y}\boldsymbol{i}_y + \frac{\partial f}{\partial z}\boldsymbol{i}_z \tag{A.1}$$

is called the **gradient** of f. In the above, $\boldsymbol{i}_x, \boldsymbol{i}_y$, and $\boldsymbol{i}_z$ are unit vectors along the x-, y-, and z-axes, respectively. Thus, grad is an operator that operates on a scalar to result in a vector. grad f points in the direction of maximum variation with a magnitude equal to the maximum variation. For example, when temperature T varies in space and κ is a heat conductivity, $-\kappa\,\text{grad } T$ gives a heat that flows across a unit area in unit time.

Using an operator,

$$\nabla = \frac{\partial}{\partial x}\boldsymbol{i}_x + \frac{\partial}{\partial y}\boldsymbol{i}_y + \frac{\partial}{\partial z}\boldsymbol{i}_z, \tag{A.2}$$

Equation (A.1) is also written as

$$\text{grad } f = \nabla f, \tag{A.3}$$

and ∇ is called **nabla**.

When $\boldsymbol{A}(x, y, z) = \left(A_x, A_y, A_z\right)$ is a given vector function,

$$\text{div } \boldsymbol{A} = \nabla \cdot \boldsymbol{A} = \frac{\partial A_x}{\partial x} + \frac{\partial A_y}{\partial y} + \frac{\partial A_z}{\partial z} \tag{A.4}$$

© The Editor(s) (if applicable) and The Author(s), under exclusive license
to Springer Nature Switzerland AG 2023
T. Matsushita, *Electricity*,
https://doi.org/10.1007/978-3-031-44002-1

is called the **divergence** of A. Thus, div is an operator that operates on a vector to result in a scalar.

When $A(x, y, z) = (A_x, A_y, A_z)$ is a given vector function,

$$
\begin{aligned}
\text{curl } A &= \nabla \times A \\
&= i_x \left(\frac{\partial A_z}{\partial y} - \frac{\partial A_y}{\partial z} \right) + i_y \left(\frac{\partial A_x}{\partial z} - \frac{\partial A_z}{\partial x} \right) + i_z \left(\frac{\partial A_y}{\partial x} - \frac{\partial A_x}{\partial y} \right)
\end{aligned}
\tag{A.5}
$$

is called the **curl** of A. Thus, curl is an operator that operates on a vector to result in a vector.

The following relations hold for various products:

$$
\text{grad}(\phi\psi) = \phi \, \text{grad } \psi + \psi \, \text{grad } \phi,
\tag{A.6}
$$

$$
\begin{aligned}
\text{grad}(A \cdot B) &= (A \cdot \nabla)B + (B \cdot \nabla)A \\
&\quad + A \times \text{curl } B + B \times \text{curl } A,
\end{aligned}
\tag{A.7}
$$

$$
\text{div}(\phi A) = \phi \, \text{div } A + \text{grad } \phi \cdot A,
\tag{A.8}
$$

$$
\text{div}(A \times B) = B \cdot \text{curl } A - A \cdot \text{curl } B,
\tag{A.9}
$$

$$
\text{curl}(\phi A) = \phi \, \text{curl } A - A \times \text{grad } \phi,
\tag{A.10}
$$

$$
\begin{aligned}
\text{curl}(A \times B) &= (B \cdot \nabla)A - (A \cdot \nabla)B \\
&\quad + A \, \text{div } B - B \, \text{div } A.
\end{aligned}
\tag{A.11}
$$

There are three formulae for second differentiation:

$$
\text{div}(\text{curl } A) = 0.
\tag{A.12}
$$

$$
\text{curl}(\text{grad } \phi) = 0.
\tag{A.13}
$$

$$
\text{curl}(\text{curl} A) = \text{grad}(\text{div } A) - \nabla^2 A.
\tag{A.14}
$$

In Cartesian coordinates, the second term is written as

$$
\nabla^2 A = \left(\frac{\partial^2}{\partial x^2} + \frac{\partial^2}{\partial y^2} + \frac{\partial^2}{\partial z^2} \right) A.
\tag{A.15}
$$

When ∇^2 operates on a scalar, it is given by

$$\nabla^2 \phi = (\nabla \cdot \nabla)\phi = \nabla \cdot (\nabla \phi). \tag{A.16}$$

It should be noted, however, when it operates on a vector, we have

$$\nabla^2 A = (\nabla \cdot \nabla)A \neq \nabla(\nabla \cdot A). \tag{A.17}$$

A.1.2 Formulae for Cylindrical and Polar Coordinates

(a) Cylindrical Coordinates

The central z-axis is defined, and a position is expressed with the distance R from the axis and the azimuthal angle φ in the plane normal to the axis, as shown in Fig. A.1. Such a coordinates is called **cylindrical coordinates**. The position is expressed as (R, φ, z). The gradient, divergence, and curl in cylindrical coordinates are

$$\text{grad } f = i_R \frac{\partial f}{\partial R} + i_\varphi \frac{1}{R} \cdot \frac{\partial f}{\partial \varphi} + i_z \frac{\partial f}{\partial z}, \tag{A.18}$$

$$\text{div } A = \frac{1}{R} \cdot \frac{\partial (R A_R)}{\partial R} + \frac{1}{R} \cdot \frac{\partial A_\varphi}{\partial \varphi} + \frac{\partial A_z}{\partial z}, \tag{A.19}$$

Fig. A.1 Cylindrical coordinates

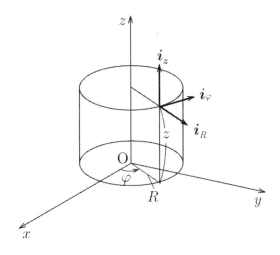

$$\text{curl } A = i_R \left(\frac{1}{R} \cdot \frac{\partial A_z}{\partial \varphi} - \frac{\partial A_\varphi}{\partial z} \right) + i_\varphi \left(\frac{\partial A_R}{\partial z} - \frac{\partial A_z}{\partial R} \right)$$

$$+ i_z \frac{1}{R} \left[\frac{\partial (R A_\varphi)}{\partial R} - \frac{\partial A_R}{\partial \varphi} \right], \tag{A.20}$$

where i_R, i_φ, and i_z are unit vectors along the radial, azimuthal, and z-axis directions, respectively, and follow the right-hand rule in the order $i_R \rightarrow i_\varphi \rightarrow i_z \rightarrow i_R$. A_R, A_φ, and A_z are the R, φ, and z-components of A.

(b) Polar Coordinates

We first define the center with an axis that determines the two poles, and then, a position is expressed with the distance r from the center, the zenithal angle θ measured from the north pole, and the azimuthal angle φ in the plane that includes the center and is normal to the axis, as shown in Fig. A.2. Such a coordinates is called **polar coordinates**. The position is expressed as (r, θ, φ). The gradient, divergence, and curl in polar coordinates are

$$\text{grad } f = i_r \frac{\partial f}{\partial r} + i_\theta \frac{1}{r} \cdot \frac{\partial f}{\partial \theta} + i_\varphi \frac{1}{r \sin \theta} \cdot \frac{\partial f}{\partial \varphi}, \tag{A.21}$$

$$\text{div } A = \frac{1}{r^2} \cdot \frac{\partial (r^2 A_r)}{\partial r} + \frac{1}{r \sin \theta} \cdot \frac{\partial (\sin \theta A_\theta)}{\partial \theta} + \frac{1}{r \sin \theta} \cdot \frac{\partial A_\varphi}{\partial \varphi}, \tag{A.22}$$

$$\text{curl } A = i_r \frac{1}{r \sin \theta} \left[\frac{\partial (\sin \theta A_\varphi)}{\partial \theta} - \frac{\partial A_\theta}{\partial \varphi} \right]$$

$$+ i_\theta \frac{1}{r} \left[\frac{1}{\sin \theta} \cdot \frac{\partial A_r}{\partial \varphi} - \frac{\partial (r A_\varphi)}{\partial r} \right]$$

Fig. A.2 Polar coordinates

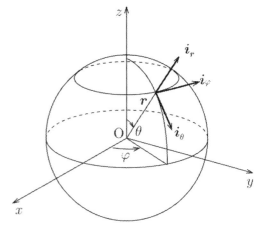

$$+ i_\varphi \frac{1}{r} \left[\frac{\partial (r A_\theta)}{\partial r} - \frac{\partial A_r}{\partial \theta} \right], \tag{A.23}$$

where i_r, i_θ, and i_φ are unit vectors along the radial, zenithal, and azimuthal directions, respectively, and follow the right-hand rule in the order $i_r \rightarrow i_\theta \rightarrow i_\varphi \rightarrow i_r$. A_r, A_θ, and A_φ are the r, θ, and φ-components of A.

A.1.3 Integral

We denote the tangential component of a vector $F(x, y, z)$ on a smooth curve C by $F_t(x, y, z)$ and the elementary line vector on C by ds, as shown in Fig. A.3. Then,

$$\int_C F(x, y, z) \cdot ds = \int_C F_t(x, y, z) ds \tag{A.24}$$

is called a **curvilinear integral**. Curvilinear integral is an inverse operation of gradient, and we have

$$\int_C \operatorname{grad} \phi \cdot ds = \int_C \frac{\partial \phi}{\partial x} dx + \int_C \frac{\partial \phi}{\partial y} dy + \int_C \frac{\partial \phi}{\partial z} dz = \int_C d\phi. \tag{A.25}$$

Hence, when C is a closed line, we have

$$\oint_C \operatorname{grad} \phi \cdot ds = 0. \tag{A.26}$$

We denote the normal component of vector $F(x, y, z)$ on a curved surface S by $F_n(x, y, z)$ and the elementary surface vector on S by dS, as shown in Fig. A.4. Then,

Fig. A.3 Curvilinear integral of vector F and its tangential component F_t on C

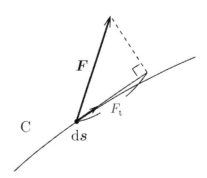

Fig. A.4 Vector **F** on
curved surface S and its
normal component F_n

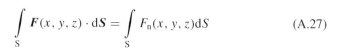

$$\int_S F(x, y, z) \cdot dS = \int_S F_n(x, y, z) dS \qquad (A.27)$$

is called a **surface integral**.

We denote a closed surface and its interior by S and V, respectively. The following relation holds for a given vector **A**:

$$\int_S A \cdot dS = \int_V \text{div } A \, dV. \qquad (A.28)$$

This is called **Gauss's theorem**. The element surface vector d**S** is directed outward, as shown in Fig. A.5, and the right side is a usual volume integral.

We denote a closed line and a surface surrounded by it by C and S, respectively. The following relation holds for a given vector **A**:

$$\oint_C A \cdot ds = \int_S \text{curl } A \cdot dS. \qquad (A.29)$$

This is called **Stokes' theorem**. The elementary surface vector d**S** points along the direction of motion of a screw when we rotate it in the direction of d**s** (see Fig. A.6).

Fig. A.5 Closed surface S
and its interior V. Elementary
surface vector d**S** is normal
to the surface and directed
outward

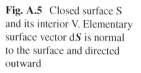
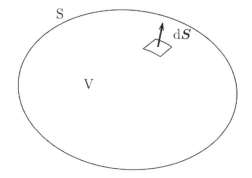

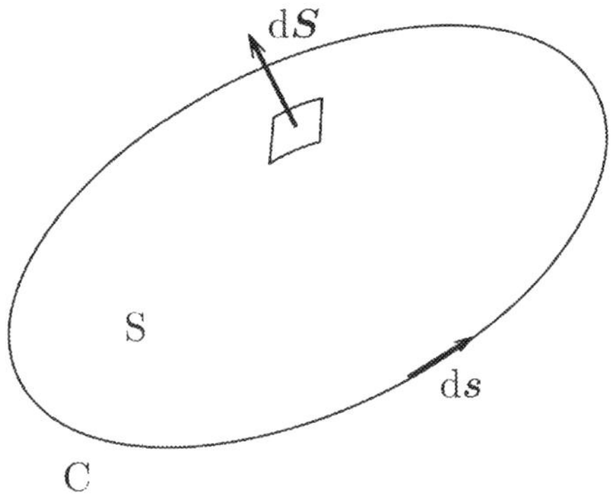

Fig. A.6 Closed line C and
surface S surrounded by it.
The elementary surface
vector dS and the elementary
line vector ds follow the
right-hand rule

A.2 Differential Forms of Laws on Electromagnetism

A.2.1 Electrostatic Field

Since the electrostatic field satisfies Eq. (1.17), it can be expresses by a gradient of
some scalar function, as shown by Eq. (A.26). In fact, it is given by Eq. (1.21) with
the electric potential ϕ. That is,

$$E = -\mathrm{grad}\,\phi. \tag{A.30}$$

Thus, from Eq. (A.13) the electrostatic field satisfies

$$\mathrm{curl}\,E = 0. \tag{A.31}$$

This shows that the electrostatic field is a field without vortex.

If we use Gauss's theorem (A.28) to the surface integral on the left side of
Eq. (1.25), it leads to

$$\int_V \mathrm{div}\,E\,\mathrm{d}V = \frac{1}{\epsilon_0}\int_V \rho\,\mathrm{d}V. \tag{A.32}$$

Since this holds for any space V, we have

$$\mathrm{div}\,E = \frac{1}{\epsilon_0}\rho. \tag{A.33}$$

Using the electric flux density, D, it leads to

$$\mathrm{div}\,D = \rho. \tag{A.34}$$

This is **Gauss's divergences law**. That is, when there are electric charges, those produce divergences of electric field and electric flux density.

A.2.2 Steady Current

We change the order of time differential and spatial integral of the left side of Eq. (3.4) and apply Gauss's theorem on the right side. Then, this equation leads to

$$\int_V \frac{\partial \rho}{\partial t} dV = -\int_V \operatorname{div} \boldsymbol{i} \, dV. \tag{A.35}$$

Since this holds for arbitrary V, we have

$$\operatorname{div} \boldsymbol{i} = -\frac{\partial \rho}{\partial t}. \tag{A.36}$$

In the case of steady current in which the electric charge density does not change with time, Eq. (A.36) leads to

$$\operatorname{div} \boldsymbol{i} = 0. \tag{A.37}$$

A.2.3 Static Magnetic Field

In a static case, the magnetic flux density satisfies Eq. (4.8). Application of Gauss's theorem to the left side leads to

$$\int_V \operatorname{div} \boldsymbol{B} \, dV = 0. \tag{A.38}$$

Thus, we have

$$\operatorname{div} \boldsymbol{B} = 0. \tag{A.39}$$

This shows that the magnetic flux density is a field without divergence. From this with Eq. (A.12), the magnetic flux density can be given by a curl of a vector:

$$\boldsymbol{B} = \operatorname{curl} \boldsymbol{A}. \tag{A.40}$$

This vector, $\boldsymbol{A}$, is called the **vector potential**.

Application of Stokes' theorem, Eq. (A.29), to the left side of Eq. (4.17) leads to

$$\int_S \text{curl } \boldsymbol{B} \cdot \text{d}\boldsymbol{S} = \mu_0 \int_S \boldsymbol{i} \cdot \text{d}\boldsymbol{S}. \tag{A.41}$$

Thus, we have

$$\text{curl } \boldsymbol{B} = \mu_0 \boldsymbol{i}. \tag{A.42}$$

If we use the magnetic field, $\boldsymbol{H}$, Eq. (A.42) leads to

$$\text{curl } \boldsymbol{H} = \boldsymbol{i}. \tag{A.43}$$

This is called a **differential form of Ampere's law**. That is, currents produce vortices of magnetic flux density and magnetic field.

A.2.4 Time-Dependent Electromagnetic Fields

Differential forms of **Maxwell's** Eqs. (6.19) and (6.20) are, respectively, given by

$$\text{curl } \boldsymbol{E} = -\frac{\partial \boldsymbol{B}}{\partial t}, \tag{A.44}$$

$$\text{curl } \boldsymbol{H} = \boldsymbol{i} + \frac{\partial \boldsymbol{D}}{\partial t}. \tag{A.45}$$

The electric potential ϕ and vector potential $\boldsymbol{A}$ that provide these time-dependent electromagnetic fields are called the **electromagnetic potential**. The corresponding equations are

$$\boldsymbol{E} = -\text{grad } \phi - \frac{\partial \boldsymbol{A}}{\partial t}, \tag{A.46}$$

$$\boldsymbol{B} = \text{curl } \boldsymbol{A}. \tag{A.47}$$

Answers to Exercises

Chapter 1

1.1 The electric field strength due to charge Q is $Q/[4\pi\epsilon_0(a^2 + d^2)]$ and is directed from Q to point A, as shown in Fig. B.1. On the other hand, the electric field strength due to charge $-Q$ is the same but is directed from point A to charge $-Q$. Thus, the electric field strength is

$$E = \frac{aQ}{2\pi\epsilon_0(a^2 + d^2)^{3/2}}$$

and is directed to right.

The electric potential at A due to charge Q is $Q/\left[4\pi\epsilon_0(a^2 + d^2)^{1/2}\right]$ and that due to $-Q$ is $-Q/\left[4\pi\epsilon_0(a^2 + d^2)^{1/2}\right]$. Thus, the total electric potential is zero. This result can also be derived using Eq. (1.14), as was done in Example 1.3. We define a straight line elongated from the midpoint between the two charges to infinity, and integrate the electric field strength from infinity to

Fig. B.1 Electric field strength due to each electric charge and combined electric field strength

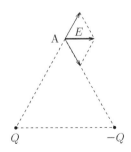

T. Matsushita, *Electricity*,
https://doi.org/10.1007/978-3-031-44002-1

point A on this line. The electric field strength is normal to the elementary line vector; $\boldsymbol{E} \cdot \mathrm{d}\boldsymbol{s} = 0$. Thus, we have $\phi = 0$.

1.2 The electric field strength due to each electric charge on the x-axis has magnitude $Q/(4\pi\epsilon_0 a^2)$ and is directed to the negative x-axis. Thus, the electric field strength due to these two charges is $Q/(2\pi\epsilon_0 a^2)$ and is directed to the negative x-axis. The electric field strength due to two electric charges on the y-axis is similarly obtained as $Q/(2\pi\epsilon_0 a^2)$ and is directed to the negative y-axis. The electric field strength due to two electric charges on the z-axis is $Q/(2\pi\epsilon_0 a^2)$ and is directed to the negative z-axis. As a result, the electric field strength at the origin is

$$E = \frac{\sqrt{3}Q}{2\pi\epsilon_0 a^2}$$

and is directed from the origin to $(x, y, z) = (-1, -1, -1)$.

The electric potential due to one positive charge is $Q/(4\pi\epsilon_0 a)$ and that due to one negative charge is $-Q/(4\pi\epsilon_0 a)$. Thus, the total electric potential is $\phi = 0$.

1.3 Since the electric field at point A produced by Q and Q_y must lie on the line connecting Q_x and point A, we have $Q_y = Q$. Then, the electric field strength produced by Q and Q_y is

$$E = \frac{\sqrt{3}Q}{4\pi\epsilon_0 a^2},$$

and the electric field strength at point A produced by Q_x is

$$E_x = \frac{Q_x}{4\pi\epsilon_0\left(\sqrt{3}a\right)^2} = \frac{Q_x}{12\pi\epsilon_0 a^2}.$$

From the requirement $E + E_x = 0$, we have $Q_x = -3\sqrt{3}Q$. The electric potential at A is

$$\phi = \frac{Q}{4\pi\epsilon_0 a} \times 2 + \frac{Q_x}{4\sqrt{3}\pi\epsilon_0 a} = -\frac{Q}{4\pi\epsilon_0 a}.$$

1.4 Suppose a plane on which the circle is placed. We define the origin at the center of the circle and the azimuthal angle φ, as shown in Fig. B.2. We regard electric charge in the region from φ to $\varphi + \mathrm{d}\varphi$, $\mathrm{d}q = a\mathrm{d}\varphi\lambda$, as a point charge. The electric field strength due to this charge on point P is $\mathrm{d}E' = a\mathrm{d}\varphi\lambda/[4\pi\epsilon_0(a^2 + b^2)]$, and only the vertical component, $\mathrm{d}E = \mathrm{d}E' \cdot b/(a^2 + b^2)^{1/2}$, remains from symmetry. Thus, the total electric field strength is given by

Fig. B.2 Electric field
strength due to electric
charge of a small region on
the circle

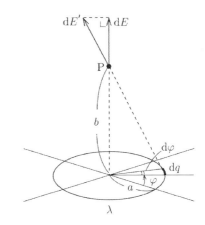

$$E = \int_0^{2\pi} \frac{a\lambda d\varphi}{4\pi\epsilon_0 (a^2 + b^2)^{3/2}} = \frac{a\lambda}{2\epsilon_0 (a^2 + b^2)^{3/2}}.$$

The electric potential due to the electric charge in the region from φ to $\varphi + d\varphi$ is given by $a\lambda d\varphi / \left[4\pi\epsilon_0 (a^2 + b^2)^{1/2} \right]$. Hence, the total electric potential is

$$\phi = \int_0^{2\pi} \frac{a\lambda d\varphi}{4\pi\epsilon_0 (a^2 + b^2)^{1/2}} = \frac{a\lambda}{2\epsilon_0 (a^2 + b^2)^{1/2}}.$$

Regarding b as a variable, we can also determine the electric field from $E = -\partial\phi/\partial b$ with this electric potential.

1.5 Suppose electric charge on a thin ring in the region from r to $r + dr$ from the center. Using the result in Exercise 1.4, the electric field strength due to this ring charge is determined to be $dE = b r \sigma dr / \left[2\epsilon_0 (r^2 + b^2)^{3/2} \right]$, where the linear charge density λ is replaced by σdr. Integrating this from 0 to R with respect to r, we obtain

$$E = \frac{b\sigma}{2\epsilon_0} \left[-\frac{1}{(r^2 + b^2)^{1/2}} \right]_0^R = \frac{b\sigma}{2\epsilon_0} \left[\frac{1}{b} - \frac{1}{(R^2 + b^2)^{1/2}} \right].$$

The electric potential due to the ring charge is given by $d\phi = r\sigma dr / \left[2\epsilon_0 (r^2 + b^2)^{1/2} \right]$. Integrating this, the electric potential is determined to be

$$\phi = \frac{\sigma}{2\epsilon_0} \left[(r^2 + b^2)^{1/2} \right]_0^R = \frac{\sigma}{2\epsilon_0} \left[(R^2 + b^2)^{1/2} - b \right].$$

In this case, we can also determine the electric field from $E = -\partial\phi/\partial b$ with this electric potential.

1.6 $E = \sigma/2\epsilon_0$ is obtained by taking a limit $R \to \infty$ in the result of Exercise 1.5. This is the same as the result of Example 1.5.

1.7 The distance of point P from the center of one side is $r = [z^2 + (a^2/4)]^{1/2}$. The result of Example 1.2 can be used to determine the electric field strength due to the electric charge on one side. In this case the upper limit of the integral is not $\pi/2$ but is $\theta' = \tan^{-1}(a/2r)$. Hence, the electric field strength due to one side is

$$E' = \frac{a\lambda}{4\pi\epsilon_0 r[r^2 + (a^2/4)]^{1/2}}.$$

The vertical component only remains without cancellation due to symmetry. From the contribution from four sides, the electric field strength is determined to be

$$E = 4E'\frac{z}{r} = \frac{a\lambda z}{\pi\epsilon_0 r^2[r^2 + (a^2/4)]^{1/2}} = \frac{a\lambda z}{\pi\epsilon_0[z^2 + (a^2/4)][z^2 + (a^2/2)]^{1/2}}.$$

1.8 We define the x-axis with the origin at the foot of the vertical line from point P on the plane as shown in Fig. B.3. Using Gauss's law, the electric field strength at point P produced by the electric charge in the region from x to $x + dx$ is $dE' = \sigma dx/[2\pi\epsilon_0(x^2 + a^2)^{1/2}]$. The vertical component only remains, and the electric field strength is given by

$$E = 2\int_0^w \frac{a\sigma dx}{2\pi\epsilon_0(x^2 + a^2)}.$$

Fig. B.3 Electric field strength and its vertical component produced by electric charge in the region from x to $x + dx$ on the plane

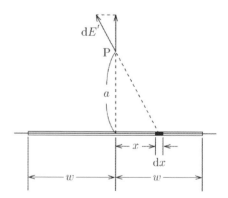

Here, we put $x = a \tan \theta$ and $\theta_w = \tan^{-1}(w/a)$. Then, the electric field strength is determined to be

$$E = 2 \int_0^{\theta_w} \frac{\sigma \, d\theta}{2\pi \epsilon_0} = \frac{\sigma \theta_w}{\pi \epsilon_0}.$$

1.9 We determine the electric field strength using Gauss's law. From symmetry, the electric field strength is $E(r) = 0$ for $0 \le r < a$ and $E(r) = a^2\sigma/(\epsilon_0 r^2)$ for $r > a$. Thus, the electric potential at the center of the sphere is determined to be

$$\phi = -\int_\infty^0 E(r) dr = -\int_\infty^a \frac{a^2\sigma}{\epsilon_0 r^2} dr = \frac{a\sigma}{\epsilon_0}.$$

The electric potential can also be directly obtained using Eq. (1.20). Since all electric charges stay at the same distance a from the observation point, we have

$$\phi = \frac{1}{4\pi \epsilon_0 a} \int \sigma \, dS = \frac{a\sigma}{\epsilon_0}.$$

1.10 We denote the distance from the central axis by R. Using Gauss's law, the electric field strength is determined to be $E(R) = 0$ for $0 \le R < a$,

$$E(R) = \frac{1}{2\pi \epsilon_0 R} \int_a^R 2\pi \rho R' dR' = \frac{\rho(R^2 - a^2)}{2\epsilon_0 R}$$

for $a < R < b$, and

$$E(R) = \frac{1}{2\pi \epsilon_0 R} \int_a^b 2\pi \rho R' dR' = \frac{\rho(b^2 - a^2)}{2\epsilon_0 R}$$

for $R > b$. The electric potential is

$$\phi(R) = \frac{\rho(b^2 - a^2)}{2\epsilon_0} \log \frac{R_\infty}{R}$$

for $b < R < R_\infty$,

$$\phi(R) = \frac{\rho b^2}{2\epsilon_0} \log \frac{R_\infty}{b} - \frac{\rho a^2}{2\epsilon_0} \log \frac{R_\infty}{R} + \frac{\rho}{4\epsilon_0}(b^2 - R^2)$$

for $a < R < b$, and

$$\phi(R) = \frac{\rho b^2}{2\epsilon_0}\log\frac{R_\infty}{b} - \frac{\rho a^2}{2\epsilon_0}\log\frac{R_\infty}{a} + \frac{\rho}{4\epsilon_0}(b^2 - a^2)$$

for $0 \le R < a$.

1.11 The electric field strength due to the electric charge that is distributed with density ρ in the whole sphere of radius b is $E_1 = \rho b^3/(3\epsilon_0 d^2)$ directed from the center to point P, and the electric potential is $\phi_1 = \rho b^3/(3\epsilon_0 d)$. The electric field strength due to the electric charge that is distributed with density $-\rho$ in the void region of radius a is $E_2 = \rho a^3/[3\epsilon_0(d-c)^2]$ directed from point P to the center, and the electric potential is $\phi_2 = -\rho a^3/[3\epsilon_0(d-c)]$. Thus, the electric field strength is

$$E = E_1 - E_2 = \frac{\rho}{3\epsilon_0}\left[\frac{b^3}{d^2} - \frac{a^3}{(d-c)^2}\right]$$

directed from the center to point P, and the electric potential is

$$\phi = \phi_1 + \phi_2 = \frac{\rho}{3\epsilon_0}\left(\frac{b^3}{d} - \frac{a^3}{d-c}\right).$$

Chapter 2

2.1 The electric charge given to the internal sphere is denoted by Q_x. The electric field strength is $E(r) = Q_x/(4\pi\epsilon_0 r^2)$ for $a < r < 2a$, $E(r) = (Q + Q_x)/(4\pi\epsilon_0 r^2)$ for $r > 3a$, and 0 for other places. Thus, the electric potential in the region $a < r < 2a$ is

$$\phi(r) = \frac{Q + Q_x}{12\pi\epsilon_0 a} + \frac{Q_x}{4\pi\epsilon_0}\left(\frac{1}{r} - \frac{1}{2a}\right).$$

Hence, from the condition, $\phi(3a/2) = 0$, we have

$$Q_x = -\frac{2Q}{3}.$$

2.2 On the surface at $r = a$, the density of electric charge is $\sigma_a = Q_1/(4\pi a^2)$, and the electric field has strength $E(a) = Q_1/(4\pi\epsilon_0 a^2)$ and is directed outward. Thus, the relationship of (2.4), $E(a) = \sigma_a/\epsilon_0$, holds. On the surface at $r = c$, the density of electric charge and electric field strength are $\sigma_c = (Q_1 + Q_2)/(4\pi c^2)$ and $E(c) = (Q_1 + Q_2)/(4\pi\epsilon_0 c^2)$, respectively. So, the same relationship holds. On the surface at $r = b$, the density of electric charge is $\sigma_b = -Q_1/(4\pi b^2)$ and the electric field strength directed inward is $E(b) = Q_1/(4\pi\epsilon_0 b^2)$. Thus, Eq. (2.4) holds taking account of the direction of the electric field and the sign of the electric charge density.

2.3 We denote by Q_x the electric charge on the right surface of the left conductor. Then, the electric charge on the left surface of this conductor is $Q - Q_x$. Thus, the total electric charge in the right side of this conductor including the right conductor is $Q_x + 2Q$. These electric charges must be equal to each other so that the electric field does not penetrate the left conductor. Thus, we have

$$Q_x = -\frac{Q}{2}.$$

Then, the electric charge on the left surface of this conductor is $3Q/2$. We denote by Q_y the electric charge on the left surface of the right conductor. Then, from the similar condition, $Q + Q_y = 2Q - Q_y$, we have $Q_y = Q/2$, and the electric charge on the left surface of this conductor is $3Q/2$. As a result, the electric charges are $3Q/2$, $-Q/2$, $Q/2$, and $3Q/2$ from the left to right surfaces. If Gauss's law is used to determine the electric charge, we need a value of electric field strength in some vacuum region.

2.4 The electric potential in vacuum$(z > 0)$ produced by the given and image charges is

$$\phi = \frac{1}{4\pi\epsilon_0} \left\{ \frac{q}{[x^2 + y^2 + (z - a)^2]^{1/2}} - \frac{q}{[x^2 + y^2 + (z + a)^2]^{1/2}} \right\}$$

and the components of electric field strength are given by

$$E_x = -\frac{\partial\phi}{\partial x} = \frac{q}{4\pi\epsilon_0} \left\{ \frac{x}{[x^2 + y^2 + (z - a)^2]^{3/2}} - \frac{x}{[x^2 + y^2 + (z + a)^2]^{3/2}} \right\},$$

$$E_y = -\frac{\partial\phi}{\partial y} = \frac{q}{4\pi\epsilon_0} \left\{ \frac{y}{[x^2 + y^2 + (z - a)^2]^{3/2}} - \frac{y}{[x^2 + y^2 + (z + a)^2]^{3/2}} \right\},$$

$$E_z = -\frac{\partial\phi}{\partial z} = \frac{q}{4\pi\epsilon_0} \left\{ \frac{z - a}{[x^2 + y^2 + (z - a)^2]^{3/2}} - \frac{z + a}{[x^2 + y^2 + (z + a)^2]^{3/2}} \right\}.$$

These components on the conductor surface $(z = 0)$ are $E_x = 0$, $E_y = 0$, and

$$E_z = -\frac{qa}{2\pi\epsilon_0(x^2 + y^2 + a^2)^{3/2}}.$$

Thus, Eq. (2.11) is derived.

2.5 Since the parallel component of the electric field strength is continuous, as shown in Eq. (2.31), the electric field strength E in the dielectric material is equal to E_0. Then, the electric flux density is given by $D = \epsilon E = \epsilon E_0$, and we have the electric polarization as $P = D - \epsilon_0 E = (\epsilon - \epsilon_0)E_0$. Since the electric polarization is parallel to the surface, the density of polarization charge on the surface is zero.

2.6 Suppose that electric charges Q and $-Q$ are given to the inner and outer electrodes, respectively. The electric flux density at r from the center is $D = Q/(4\pi r^2)$ in both dielectric materials. So, the electric field strength is $E = Q/(4\pi \epsilon_1 r^2)$ for $a < r < b$ and $E = Q/(4\pi \epsilon_2 r^2)$ for $b < r < c$. The electric potential difference between the two electrodes is

$$V = \int_a^b \frac{Q}{4\pi \epsilon_1 r^2} dr + \int_b^c \frac{Q}{4\pi \epsilon_2 r^2} dr = \frac{(b-a)Q}{4\pi \epsilon_1 ab} + \frac{(c-b)Q}{4\pi \epsilon_2 bc}.$$

Thus, the capacitance is given by

$$C = \frac{4\pi abc \epsilon_1 \epsilon_2}{\epsilon_1 a(c-b) + \epsilon_2 c(b-a)}.$$

2.7 Suppose that electric charges Q and $-Q$ are given to the inner and outer electrodes, respectively. The electric charge density on the surface of the inner electrode is not uniform, and its values on the surface of the inner electrode facing to dielectric material 1 (ϵ_1) and dielectric material 2 (ϵ_2) are denoted by σ_1 and σ_2, respectively. So, we have

$$Q = 2\pi a^2 (\sigma_1 + \sigma_2).$$

The electric field strength at distance r from the center in each dielectric material is $E_1 = \sigma_1 a^2/(\epsilon_1 r^2)$ and $E_2 = \sigma_2 a^2/(\epsilon_2 r^2)$. Since the electric potential difference between the electrodes must be the same in each region, the following condition must be satisfied:

$$V = \frac{\sigma_1 a(b-a)}{\epsilon_1 b} = \frac{\sigma_2 a(b-a)}{\epsilon_2 b}.$$

Hence, the total electric charge is given by $Q = 2\pi ab(\epsilon_1 + \epsilon_2)V/(b-a)$, and we have

$$C = \frac{2\pi ab(\epsilon_1 + \epsilon_2)}{b-a}.$$

2.8 We denote the distance from the center of the conductor by r. The electric field strength is 0 for $0 \le r < a$ and $Q/(4\pi \epsilon_0 r^2)$ for $r > a$. So, the electrostatic energy density is 0 for $0 \le r < a$ and $Q^2/(32\pi^2 \epsilon_0 r^4)$ for $r > a$. Then, the electrostatic energy is determined to be

$$U_e = \int_a^\infty \frac{Q^2}{32\pi^2 \epsilon_0 r^4} 4\pi r^2 dr = \frac{Q^2}{8\pi \epsilon_0 a}.$$

2.9   Since the electric potential of the spherical conductor is $\phi = Q/(4\pi\epsilon_0 a)$, the electrostatic energy is given by

$$U_e = \frac{1}{2} Q\phi = \frac{Q^2}{8\pi\epsilon_0 a},$$

which agrees to the result of 2.8.

2.10 Suppose that the voltage between the two electrodes is V, when the dielectric plate is inserted by distance x from the edge. In this case, the electric field strength is $E = V/d$ in both of vacuum and dielectric plate. Hence, the electric flux density is $D_0 = \epsilon_0 V/d$ and $D = \epsilon V/d$ in the respective regions. Since the density of electric charge that appears on the electrode surface facing vacuum and the dielectric material is $\sigma_0 = D_0$ and $\sigma = D$, respectively, the total electric charge is given by

$$Q = [\sigma_0(a - x) + \sigma x]b = \frac{[\epsilon_0(a - x) + \epsilon x]bV}{d}.$$

Thus, V is determined by this condition. Hence, using Q, the electrostatic energy is written as

$$U_e = \frac{1}{2} QV = \frac{dQ^2}{2b[\epsilon_0 a + (\epsilon - \epsilon_0)x]}.$$

Note that V changes with x. The force on the dielectric plate is determined to be

$$F = -\frac{\partial U_e}{\partial x} = \frac{d(\epsilon - \epsilon_0)Q^2}{2b[\epsilon_0 a + (\epsilon - \epsilon_0)x]^2}.$$

Since $\epsilon > \epsilon_0$, we have $F > 0$. This is directed to the increasing x, so the attractive force works.

Chapter 3

3.1   The electric charge stored in the capacitor is $Q = 3 \times 10^{-6} \times 20 = 6 \times 10^{-5}$ C. So, the voltage between the terminal is $V = Q/C = 6 \times 10^{-5}/1 \times 10^{-6} = 60$ V.

3.2   We put $b = a(1 + \delta)$. If δ is small enough, we have $\log(b/a) = \log(1 + \delta) \simeq \delta - \delta^2/2$. In the limit $\delta \to 0$, the resistance approaches

$$R \to \frac{\rho_r l}{ta},$$

which agrees with Ohm's law (3.7).

3.3   We put $d/2r_0 = \delta$. If δ is small enough, we have

$$\log \frac{r_0 + d/2}{r_0 - d/2} = \log(1 + \delta) - \log(1 - \delta) \simeq 2\delta.$$

Thus, the resistance is given by

$$R = \frac{\pi \rho_r}{4w\delta} = \frac{\pi r_0 \rho_r}{2wd},$$

which agrees with Ohm's law (3.7).

3.4 Suppose that current I is applied along the length. The radius of a part of distance x from the bottom is $r(x) = b - (b - a)x/h$, and the current density and electric field strength at this position are $I/(\pi r^2)$ and $\rho_r I/(\pi r^2)$, respectively. Thus, the voltage between the two edges is

$$V = \int_0^h \frac{\rho_r I}{\pi [b - (b - a)x/h]^2} dx = \frac{\rho_r h I}{\pi (b - a)} \left(\frac{1}{a} - \frac{1}{b} \right) = \frac{\rho_r h I}{\pi ab}.$$

Then, the resistance is determined to be

$$R = \frac{\rho_r h}{\pi ab}.$$

3.5 Suppose that current I is applied between the terminals. The current density and electric field strength at a position at a distance r from the center are $I/(4\pi r^2)$ and $\rho_r I/(4\pi r^2)$, respectively. Then, the voltage between the terminals is

$$V = \int_a^b \frac{\rho_r I}{4\pi r^2} dr = \frac{\rho_r (b - a) I}{4\pi ab},$$

and the resistance is determined to be

$$R = \frac{\rho_r (b - a)}{4\pi ab}.$$

3.6 From Eq. (3.24), the dissipated power density is

$$p = \frac{4V^2}{\pi^2 \rho_r r^2}.$$

So, the dissipated power density at the most inner part is larger by a factor;

$$\left(\frac{r_0 + d/2}{r_0 - d/2} \right)^2,$$

in comparison with the most outer part.

3.7 From Example 3.3, the dissipated power density at a position of radius r is

$$p = iE = \frac{4V^2}{\pi^2 \rho_r r^2}.$$

The total length of the region of radius from r to $r + dr$ is $\pi r/2$, and the dissipated power in this region is $dP = (\pi r w/2)p\,dr = 2wV^2 dr/(\pi \rho_r r)$. So, the total dissipated power is given by

$$P = \frac{2wV^2}{\pi \rho_r} \int_{r_0-d/2}^{r_0+d/2} \frac{dr}{r} = \frac{2wV^2}{\pi \rho_r} \log \frac{r_0 + d/2}{r_0 - d/2},$$

which agrees with the result $P = IV$.

3.8 We denote the flowing current by I. The radius of a part of height x from the bottom is given by $r(x) = b - (b-a)x/h$. So, the current density and dissipated power density at this position are given by $I/(\pi r^2)$ and $p = \rho_r I^2/(\pi r^2)^2$, respectively. Using the result of the voltage $V = \rho_r h I/\pi ab$ in Exercise 3.4, the dissipated power density is rewritten as $p = (ab)^2 V^2/(\rho_r h^2 r^4)$. Thus, the total dissipated power is determined to be

$$P = \frac{(ab)^2 V^2}{\rho_r h^2} \int_0^h \frac{1}{r^4(x)} \pi r^2(x)\,dx$$

$$= \frac{\pi (ab)^2 V^2}{\rho_r h^2} \int_0^h \frac{dx}{[b - (b-a)x/h]^2} = \frac{\pi ab V^2}{\rho_r h}.$$

Using the resistance $R = \rho_r h/\pi ab$, this result is written as $P = V^2/R$.

Chapter 4

4.1 The magnetic flux density at O produced by each current is given by $B = \mu_0 I/(2\sqrt{2}\pi a)$. But the magnetic flux density produced by pair of diagonal two currents is zero because of cancellation. So, the magnetic flux density is zero.

4.2 Since the magnetic flux density at point A produced by I and I_y must be normal to the line connecting I_x and point A, we have $I_y = I$. Then, the magnetic flux density at A produced by I and I_y is

$$B = \frac{\sqrt{3}\mu_0 I}{2\pi a},$$

and the magnetic flux density at A produced by I_x is

Fig. B.4 Combined force
produced by two currents

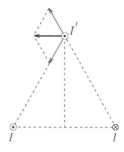

$$B_x = \frac{\mu_0 I_x}{2\sqrt{3}\pi a}.$$

From the requirement $B + B_x = 0$, we have $I_x = -3I$.

4.3 From the magnetic flux density obtained using Ampere's law, I' receives an attractive force of magnitude $\mu_0 I I'/\left[2\pi \left(a^2 + d^2\right)^{1/2}\right]$ in a unit length from the current of the same direction and a repulsive force of the same magnitude from the current of the opposite direction. As a result, the combined force is $\mu_0 a I I'/\left[\pi \left(a^2 + d^2\right)\right]$ in a unit length and is directed to left, as shown in Fig. B.4.

4.4 Since the angle θ to the center O in Eq. (4.5) is 0 and π on the two straight sections, the magnetic flux density at O from these sections is zero. The θ is $\pi/2$ on the semicircle and the magnetic flux density produced by this section is obtained from Eq. (4.5) as

$$B = \frac{\mu_0 I}{4\pi a^2} \int dr' = \frac{\mu_0 I}{4a}.$$

This is the magnetic flux density at O. The direction is backward normal to the page.

4.5 We define the origin at the center of the circle and the azimuthal angle φ, as shown in Fig. B.5. The magnetic flux density at point P produced

Fig. B.5 Magnetic flux
density produced by an
elementary current on the
circle

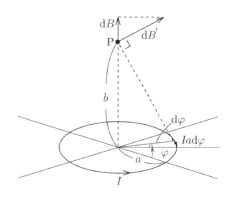

by an elementary current $I a \mathrm{d}\varphi$ in the small region from φ to $\varphi + \mathrm{d}\varphi$ is $\mathrm{d}B' = \mu_0 I a \mathrm{d}\varphi / [4\pi (a^2 + b^2)]$. Only the vertical component, $\mathrm{d}B = \mathrm{d}B' a / (a^2 + b^2)^{1/2}$, remains without cancellation. So, the total magnetic flux density is determined to be

$$B = \int_0^{2\pi} \frac{\mu_0 a^2 I \mathrm{d}\varphi}{4\pi (a^2 + b^2)^{3/2}} = \frac{\mu_0 a^2 I}{2(a^2 + b^2)^{3/2}}.$$

It is directed upward.

4.6 First, we treat the contribution from the left side. The angle from an elementary line ds directed to observation point A is denoted by θ. Then, the distance from the elementary line to A is $r = a / \sin\theta$ and $|ds| = r\mathrm{d}\theta / \sin\theta = a\mathrm{d}\theta / \sin^2\theta$. The magnetic flux density due the elementary current is

$$\mathrm{d}B = \frac{\mu_0 I \mathrm{d}s}{4\pi r^2} \sin\theta = \frac{\mu_0 I}{4\pi a} \sin\theta \mathrm{d}\theta.$$

Thus, the contribution from the left side is

$$B_1 = \frac{\mu_0 I}{4\pi a} \int_{\pi/4}^{\pi/2} \sin\theta \mathrm{d}\theta = \frac{\mu_0 I}{4\sqrt{2}\pi a}.$$

Its direction is backward normal to the page. The contribution from the bottom side is the same.

Next, we treat the contribution from the inclined side. The distance between the line and A is $a / \sqrt{2}$, and the angle θ varies from $-3\pi/4$ to $-\pi/4$. Then, the magnetic flux density is directed forward normal to the page, and its magnitude is

$$B_2 = -\frac{\mu_0 I}{2\sqrt{2}\pi a} \int_{-3\pi/4}^{-\pi/4} \sin\theta \mathrm{d}\theta = \frac{\mu_0 I}{2\pi a}.$$

Thus, the magnetic flux density is directed forward normal to the page and its magnitude is

$$B = -2B_1 + B_2 = \frac{\left(2 - \sqrt{2}\right)\mu_0 I}{4\pi a}.$$

4.7 The distance between the center of a side and point P is $r = [b^2 + (a^2/4)]^{1/2}$. We use the result of Example 4.1 to determine the magnetic flux density produced by the current flowing the side. In this case the range of θ for the integration is from θ' to $\pi - \theta'$, where $\theta' = \tan^{-1}(2r/a)$. Hence, the magnetic flux density produced by one side is

$$B' = \frac{\mu_0 I}{4\pi r} \int_{\theta'}^{\pi - \theta'} \sin\theta \, d\theta = \frac{\mu_0 I}{2\pi r} \cos\theta' = \frac{\mu_0 a I}{4\pi r [r^2 + (a^2/4)]^{1/2}}.$$

Only the vertical component remains without cancellation. From contribution from 4 sides, we have

$$B = 4B' \frac{a}{2r} = \frac{\mu_0 a^2 I}{2\pi r^2 [r^2 + (a^2/4)]^{1/2}} = \frac{\mu_0 a^2 I}{2\pi [b^2 + (a^2/4)][b^2 + (a^2/2)]^{1/2}}.$$

4.8 We define the x-axis with the origin at the foot of the vertical line from point P on the plane, as shown in Fig. B.6. Using Ampere's law, the magnetic flux density at point P produced by the current $(I/2w)dx$ flowing in a thin region from x to $x + dx$ is $dB' = \mu_0 I dx / \left[4\pi w (x^2 + a^2)^{1/2} \right]$. Only the x-component remains without cancellation, and integrating this, we have

$$B = 2 \int_0^w \frac{\mu_0 I a dx}{4\pi w (x^2 + a^2)}.$$

We put $x = a\tan\theta$. The magnetic flux density is determined to be

$$B = 2 \int_0^{\theta_w} \frac{\mu_0 I d\theta}{4\pi w} = \frac{\mu_0 I \theta_w}{2\pi w},$$

where $\theta_w = \tan^{-1}(w/a)$.

Fig. B.6 Magnetic flux density and its horizontal component produced by the current in the region from x to $x + dx$

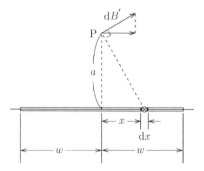

4.9 We denote the distance from the center by r. Then, using Ampere's law, we have

$$B(r) = 0; \qquad\qquad 0 \leq r < a,$$
$$= \frac{\mu_0(r^2 - a^2)I}{2\pi(b^2 - a^2)r}; \quad a < r < b,$$
$$= \frac{\mu_0 I}{2\pi r}; \qquad\qquad r > b.$$

4.10 The magnetic flux density at point P due to the current with density i in the whole cylinder of radius b is $B_1 = \mu_0 b^2 i/(2d)$ directed upward, and the magnetic flux density due to the current with density $-\rho$ in the whole cylinder of radius a is $B_2 = \mu_0 a^2 i/[2(d - c)]$ directed downward. Thus, the magnetic flux density is directed upward and the magnitude is

$$B = B_1 - B_2 = \frac{\mu_0 i}{2}\left(\frac{b^2}{d} - \frac{a^2}{d - c}\right).$$

Chapter 5

5.1 Current I flows uniformly on the surface at $r = a$ of the inner superconductor so that magnetic flux does not penetrate it. Then, current $-I$ flows uniformly on the inner surface at $r = b$ of the outer superconductor so that magnetic flux does not penetrate it. Since the total current is zero in the outer superconductor, current I flows uniformly on the outer surface at $r = c$ of the outer superconductor. Under this current distribution, the magnetic flux density in each region is determined to be

$$B(r) = 0; \qquad\qquad 0 \leq r < a,$$
$$= \frac{\mu_0 I}{2\pi r}; \qquad\qquad a < r < b,$$
$$= 0; \qquad\qquad b < r < c,$$
$$= \frac{\mu_0 I}{2\pi r}; \qquad\qquad r > c.$$

5.2 We denote by I_x the current on the right surface of the left superconductor. Then, the current on the left surface of this superconductor is $I - I_x$. Thus, the total current in the right side of this superconductor including the right superconductor is $I_x + 2I$. These currents must be equal to each other so that the magnetic flux does not penetrate the left superconductor. Thus, we have

$$I_x = -\frac{I}{2}.$$

Then, the current on the left surface of this superconductor is $3I/2$. We denote by I_y the current on the left surface of the right superconductor. Then, from the similar condition, $I + I_y = 2I - I_y$, we have $I_y = I/2$, and the current on the left surface of this superconductor is $3I/2$. As a result, the currents are $3I/2$, $-I/2$, $I/2$, and $3I/2$ from the left to right surfaces. If Ampere's law is used to determine the current, we need a value of magnetic flux density in some vacuum region.

5.3   Since the normal component of the magnetic flux density is continuous, as shown in Eq. (5.32), the magnetic flux density B in the magnetic material is equal to B_0. Then, the magnetic field strength is given by $H = B/\mu = B_0/\mu$, and we have the magnetization as

$$M = \frac{B}{\mu_0} - H = \frac{(\mu - \mu_0)}{\mu_0 \mu} B_0.$$

Since the magnetic flux density is normal to the surface, the density of magnetizing current on the surface is zero.

5.4   When we apply current I to the parallel-plate transmission line, the magnetic flux density produced in the space between the plates is $B = \mu_0 I/b$, and the total magnetic flux is $\Phi = Bad = \mu_0 Iad/b$. Thus, the inductance is determined to be

$$L = \frac{\Phi}{I} = \frac{\mu_0 ad}{b}.$$

5.5   The magnetic flux density at position z on the central axis is

$$B(z) = \frac{\mu_0 I a^2}{2} \left\{ \frac{1}{[(z+d)^2 + a^2]^{3/2}} + \frac{1}{[(z-d)^2 + a^2]^{3/2}} \right\}.$$

Thus, we have

$$\frac{\partial^2 B(z)}{\partial z^2} = \frac{3\mu_0 I a^2}{2} \left\{ \frac{4(z+d)^2 - a^2}{[(z+d)^2 + a^2]^{7/2}} + \frac{4(z-d)^2 - a^2}{[(z-d)^2 + a^2]^{7/2}} \right\}.$$

From the condition that this is equal to zero at $z = 0$, we obtain $2d = a$.

5.6   We determine the magnetic flux penetrating line 2 when we apply current I_1 to line 1. The magnetic flux produced by the right current of line 1 is directed opposite to the magnetic flux produced by the current flowing in itself. The magnetic flux at distance from b to $(b^2 + c^2)^{1/2}$ from the current penetrates line 2, as shown in Fig. B.7. So, the penetrating magnetic flux in a unit length is

Fig. B.7 Magnetic flux
penetrating line 2 produced
by the right current in line 1

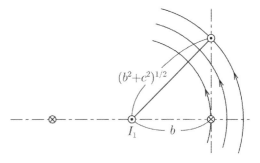

$$\Phi'_{2r} = \frac{\mu_0 I_1}{2\pi} \int_b^{(b^2+c^2)^{1/2}} \frac{dr}{r} = \frac{\mu_0 I_1}{4\pi} \log \frac{b^2 + c^2}{b^2}.$$

The magnetic flux produced by the left current of line 1 is in the same direction as the magnetic flux produced by the current flowing in itself. Its value in a unit length is

$$\Phi'_{21} = \frac{\mu_0 I_1}{2\pi} \int_{a+b}^{[(a+b)^2+c^2]^{1/2}} \frac{dr}{r} = \frac{\mu_0 I_1}{4\pi} \log \frac{(a+b)^2 + c^2}{(a+b)^2}.$$

Thus, the magnetic flux in a unit length in the designated direction is

$$\Phi'_2 = -\Phi'_{2r} + \Phi'_{21} = -\frac{\mu_0 I_1}{4\pi} \log \frac{(a+b)^2(b^2+c^2)}{b^2[(a+b)^2+c^2]},$$

and the mutual inductance in a unit length is determined to be

$$L'_{21} = \frac{\Phi'_2}{I_1} = -\frac{\mu_0}{4\pi} \log \frac{(a+b)^2(b^2+c^2)}{b^2[(a+b)^2+c^2]}.$$

5.7   When we apply current I to the left line as shown in the arrow, the magnetic flux penetrating the rectangular circuit is directed downward and its direction is the same as that produced by the current flowing in the rectangular circuit. Its value is

$$\Phi_1 = b\frac{\mu_0 I}{2\pi} \int_{d-a}^{d+a} \frac{dr}{r} = \frac{\mu_0 I b}{2\pi} \log \frac{d+a}{d-a}.$$

The magnetic flux produced by the right line is in the same direction and the magnitude is the same. Hence, the total magnetic flux is $\Phi = 2\Phi_1$ and the mutual inductance is determined to be

$$M = \frac{\Phi}{I} = \frac{\mu_0 b}{\pi} \log \frac{d+a}{d-a}.$$

5.8 The magnetic field and magnetic flux density are parallel to the plates. We denote these values in magnetic materials 1 and 2 by H_1, B_1, H_2, and B_2, respectively. Then, Ampere's law shows $H_1 = H_2 = I/w$. Hence, we have $B_1 = \mu_1 I/w$ and $B_2 = \mu_2 I/w$, and the total magnetic flux is

$$\Phi = \frac{Id(B_1 + B_2)}{2} = \frac{(\mu_1 + \mu_2)Ild}{2w}.$$

The self-inductance is determined to be

$$L = \frac{\Phi}{I} = \frac{(\mu_1 + \mu_2)ld}{2w}.$$

5.9 When we apply currents I and $-I$ to the inner and outer superconductors, magnetic flux appears only in the space between the superconductors, and the magnetic field strength at r from the center is $H(r) = I/(2\pi r)$. Then, the magnetic flux density is $B_1(r) = \mu_1 I/(2\pi r)$ and $B_2(r) = \mu_2 I/(2\pi r)$ in magnetic material 1 and 2, respectively. So, the magnetic flux in a unit length is given by

$$\Phi' = \frac{\mu_1 I}{2\pi} \int_a^b \frac{dr}{r} + \frac{\mu_2 I}{2\pi} \int_b^c \frac{dr}{r} = \frac{I}{2\pi}\left(\mu_1 \log \frac{b}{a} + \mu_2 \log \frac{c}{b}\right).$$

The self-inductance of a unit length is determined to be

$$L' = \frac{\Phi'}{I} = \frac{1}{2\pi}\left(\mu_1 \log \frac{b}{a} + \mu_2 \log \frac{c}{b}\right).$$

5.10 When we apply currents I and $-I$ to the inner and outer superconductors, respectively, the magnetic flux appears only in the space between the superconductors. Since the magnetic flux density is the same between the two magnetic materials, the current density on the superconductor surface is different between the areas facing each magnetic material. The current densities on the inner superconductor surface at $r = a$ facing magnetic materials 1 and 2 are denoted by τ_1 and τ_2, respectively. Then, the total current is given by

$$\pi a(\tau_1 + \tau_2) = I.$$

The magnetic field at $r(a < r < b)$ from the central axis is $a\tau_1/r$ in magnetic material 1 and $a\tau_2/r$ in magnetic material 2. So, the magnetic flux density is $B_1 = \mu_1 a\tau_1/r$ in magnetic material 1 and $B_2 = \mu_2 a\tau_2/r$ in magnetic material

2. The boundary condition of Eq. (5.32) requires $B_1 = B_2$, and we have $\mu_1 \tau_1 = \mu_2 \tau_2$. Thus, from the relation on the total current, the current densities are determined to be $\tau_1 = \mu_2 I / [\pi (\mu_1 + \mu_2) a]$ and $\tau_2 = \mu_1 I / [\pi (\mu_1 + \mu_2) a]$. Since the magnetic flux density is

$$B(r) = \frac{\mu_1 \mu_2 I}{\pi (\mu_1 + \mu_2) r},$$

the magnetic flux in a unit length is given by

$$\Phi' = \frac{\mu_1 \mu_2 I}{\pi (\mu_1 + \mu_2)} \int_a^b \frac{dr}{r} = \frac{\mu_1 \mu_2 I}{\pi (\mu_1 + \mu_2)} \log \frac{b}{a}.$$

The self-inductance in a unit length is determined to be

$$L' = \frac{\Phi'}{I} = \frac{\mu_1 \mu_2}{\pi (\mu_1 + \mu_2)} \log \frac{b}{a}.$$

Chapter 6

6.1 The direction of the magnetic flux produced by a current flowing as shown by the arrow is opposite to the direction of the magnetic flux when a current flows in the rectangular circuit along ABCD. The magnetic flux density at a point of distance x from the linear current is $B = -\mu_0 I / (2\pi x)$. So, the magnetic flux penetrating the rectangular circuit is

$$\Phi = -\frac{\mu_0 I b}{2\pi} \int_d^{d+a} \frac{dx}{x} = -\frac{\mu_0 I b}{2\pi} \log \frac{d+a}{d}.$$

Hence, the electromotive force is determined to be

$$V_{em} = -\frac{d\Phi}{dt} = -\frac{\partial \Phi}{\partial d} \cdot \frac{dd}{dt} = -\frac{\mu_0 I a b v}{2\pi d (d+a)}.$$

6.2 The magnetic flux density is $B(r) = \mu_1 \mu_2 I / [\pi (\mu_1 + \mu_2) r]$ in the region $a \le r \le b$ with r denoting the distance from the central axis for both magnetic materials, and is zero in other regions. Hence, the magnetic energy density is $1/(2\mu_1) B^2(r)$ and $1/(2\mu_2) B^2(r)$ in magnetic materials 1 and 2, respectively. Then, the magnetic energy in a unit length is determined to be

$$U'_m = \frac{\mu_1^2 \mu_2^2 I^2}{2\pi^2 (\mu_1 + \mu_2)^2} \int_a^b \left(\frac{1}{\mu_1} + \frac{1}{\mu_2} \right) \frac{\pi r}{r^2} dr = \frac{\mu_1 \mu_2 I^2}{2\pi (\mu_1 + \mu_2)} \log \frac{b}{a}.$$

The magnetic flux in the coaxial transmission line in a unit length is

$$\Phi' = \int_a^b B(r)dr = \frac{\mu_1\mu_2 I}{\pi(\mu_1 + \mu_2)} \log \frac{b}{a}.$$

So, the magnetic energy can also be determined from $U'_m = \Phi' I/2$.

6.3 Suppose a circle of radius r on a plane between the two electrodes with the center on the line connecting the centers of the two electrodes. We apply Eq. (6.20) on this circle. From symmetry, the magnetic field $H(r)$ is parallel to this circle and is uniform. So, the left side of Eq. (6.20) is $2\pi r H(r)$. In the region between the two electrodes, only the displacement current obtained in Example 6.5 exists. So, the right side of Eq. (6.20) is $\left(I_0 r^2/a^2\right) \sin \omega t$. Thus, the magnetic field strength is determined to be

$$H(r) = \frac{I_0 r}{2\pi a^2} \sin \omega t.$$

6.4 Differentiating Eq. (6.35) with respect to x, we have $\partial^2 B/\partial x^2 = -\mu\sigma_c \partial E/\partial x$. Substitution of Eq. (6.33) into the right side yields

$$\frac{\partial^2 B}{\partial x^2} = \mu\sigma_c \frac{\partial B}{\partial t}.$$

Thus, we have a diffusion equation on the magnetic flux density.

6.5 The dissipated power density is given by $P = \sigma_c E^2 = \sigma_c E_0^2 e^{-2x/\delta} \cos^2[\omega t - (x/\delta)]$. Integrating this with respect to time over a period $(T = 2\pi/\omega)$, we have

$$W = \frac{\sigma_c E_0^2}{2} e^{-2x/\delta}.$$

Integrating this with respect to x, the energy dissipated in a unit surface area is determined to be

$$W' = \frac{\sigma_c E_0^2}{2} \int_0^\infty e^{-2x/\delta} dx = \frac{\sigma_c \delta E_0^2}{4} = \left(\frac{\sigma_c}{8\omega\mu}\right)^{1/2} E_0^2.$$

6.6 Differentiating Eq. (6.47) with respect to x, and exchanging the order of differentiation of E on time and space for the right side, we have $\partial^2 B/\partial x^2 = -\epsilon_0\mu_0(\partial/\partial t)(\partial E/\partial x)$. Substitution of Eq. (6.33) to the right side yields

$$\frac{\partial^2 B}{\partial x^2} = \epsilon_0\mu_0 \frac{\partial^2 B}{\partial t^2}.$$

Thus, we have a wave equation on the magnetic flux density.

6.7 We average the square of the real part of the electric field of Eq. (6.50) with respect to time:

$$\langle E^2(x, t)_t \rangle = \frac{1}{T} \int_0^T \left[E_1^2 \cos^2(\omega t + kx) + 2E_1 E_2 \cos(\omega t + kx) \cos(\omega t - kx) \right.$$
$$\left. + E_2^2 \cos^2(\omega t - kx) \right] dt$$
$$= \frac{E_1^2 + E_2^2}{2} + E_1 E_2 \cos 2kx.$$

Taking average this over a wave length, the second term becomes zero, and we obtain the average of the electric energy density,

$$\langle u_e \rangle_{x,t} = \frac{\epsilon_0}{4} \left(E_1^2 + E_2^2 \right).$$

We average the square of the real part of the magnetic flux density of Eq. (6.55) with respect to time:

$$\langle B^2(x, t)_t \rangle = \frac{1}{T} \int_0^T \left[B_1^2 \cos^2(\omega t + kx) - 2B_1 B_2 \cos(\omega t + kx) \cos(\omega t - kx) \right.$$
$$\left. + B_2^2 \cos^2(\omega t - kx) \right] dt$$
$$= \frac{B_1^2 + B_2^2}{2} - B_1 B_2 \cos 2kx.$$

Taking average this over a wave length, the second term becomes zero, and we obtain the average of the magnetic energy density,

$$\langle u_m \rangle_{x,t} = \frac{1}{4\mu_0} \left(B_1^2 + B_2^2 \right) = \frac{\epsilon_0}{4} \left(E_1^2 + E_2^2 \right),$$

where $B_1 = E_1/c$, $B_2 = E_1/c$, and Eq. (6.53) are used. Thus, the magnitude is the same between the electric energy density and the magnetic energy density.

Chapter 7

7.1 The left group is a parallel connection of 6 Ω, 12 Ω, and 4 Ω. So, the combined resistance is 2 Ω. The resistance of the right group is 2 Ω. The total combined resistance is 4 Ω.

7.2 The resistance before the exchange is $R_1 = 4R/(R+4) + 12[\Omega]$, and the resistance after the exchange is $R_2 = R + 3[\Omega]$. The required condition is $R_2 = R_1/2$. This leads to

$$(R - 4)(R + 3) = 0.$$

Because of $R > 0$, we have the solution, $R = 4[\Omega]$.

7.3 The combined resistance is

$$\frac{6(R + 6)}{R + 12} + R.$$

Since this is equal to $10\ \Omega$, we have

$$(R + 14)(R - 6) = 0.$$

Because of $R > 0$, we have the solution, $R = 6[\Omega]$.

7.4 The currents I_1 and I_2 satisfy $I_1 = R_1 I/(R + R_1)$ and $I_2 = R_2 I/(R + R_2)$, respectively. Thus, the given condition is

$$\frac{R_1}{R + R_1} = \frac{2R_2}{R + R_2},$$

and we have

$$R = \frac{R_1 R_2}{R_1 - 2R_2}.$$

So that R is positive, the following condition is necessary:

$$R_1 > 2R_2.$$

7.5 From the symmetry, the resistance is equal to that shown in Fig. B.8. Thus, the combined resistance is $(13/7)R$.

7.6 We define currents $I_1 - I_6$, as shown in Fig. B.9. The total current is $2I_1$, and the combined resistance is determined, if the potential difference between the two terminals can be obtained as a function of I_1. Since I_4 is equal to I_5, we have $I_2 = 2I_4$. From the symmetry, there is no current that flows between the upper and lower halves of the circuit. So, I_3 is equal to I_6. As a result, the potential difference between A and B is given by $2(I_2 + I_4)R = 3I_2R$ or $4I_3R$. This yields $I_3 = (3/4)I_2$. Thus, we obtain $I_2 = (4/7)I_1$ and $I_3 = (3/7)I_1$. The

Fig. B.8 Equivalent circuit

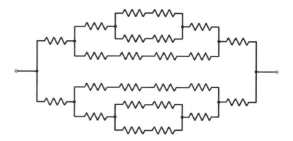

Fig. B.9 Currents flowing
through each resistor

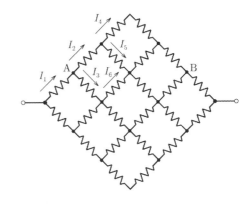

potential difference between the two terminals is given by

$$V = 2I_1 R + 4I_3 R = \frac{26}{7} I_1 R.$$

Thus, the combined resistance is determined to be $V/2I_1 = (13/7)R$.

7.7 Suppose that a voltage is applied between A and B. Since the current flowing
AO and that flowing OB are the same, the situation does not change even if
we disconnect the upper and lower parts at O. Hence, three resistors of $R, 2R$,
and $(8/3)R$ are connected in series, and the resistance between A and B is
$(8/15)R$.

Suppose that a voltage is applied between A and C. Since current does not
flow between O and B, the situation does not change even if we disconnect
O and B. We can also disconnect O and D for the same reason. Then, three
resistors of $2R$ are connected in parallel, and the resistance between A and C
is $(2/3)R$.

Suppose that a voltage is applied between O and A. Since points B and
D are equipotential, the situation does not change even if we connect these
points. Then, the resistors of $(1/2)R$ and $(3/2)R$ are connected in parallel
between O and D. Since the resistance between O and A through D is $(7/8)R$,
the resistance between O and A is $(7/15)R$.

7.8 We denote by R' the resistance on the right side of the broken line, as shown in
Fig. B.10a. Then, the total resistance can be obtained, as shown in Fig. B.10b.
The resistance R' can be obtained with the method shown in Example 7.6 as

$$R' = 2R_0 + \frac{R_0 R'}{R_0 + R'}.$$

Thus, we have $R' = \left(1 + \sqrt{3}\right) R_0$. The resistance between A and B is
determined to be

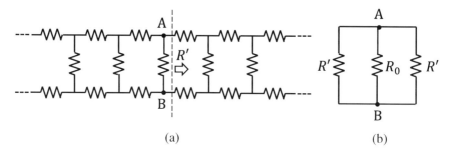

Fig. B.10 a Resistor circuit and **b** equivalent circuit

$$R = \frac{R_0 R'/2}{R_0 + (R'/2)} = \frac{R_0}{\sqrt{3}}.$$

7.9 The current flowing in the resistor of R is $I = R_1 J/(R + R_1 + R_2)$. So, the power dissipated in the resistor of R is given by

$$P = RI^2 = \frac{R_1^2 R}{(R + R_1 + R_2)^2}.$$

From the condition $\partial P/\partial R = 0$, we have

$$R = R_1 + R_2.$$

Since $\partial^2 P/\partial R^2 = -R_1^2 J^2/[8(R_1 + R_2)^3] < 0$, the above condition is the solution.

7.10 The total resistance is

$$R = R_1 + \frac{R_0 R_2}{R_0 + R_2},$$

and the total current is $I = E/R$. Hence, the current flowing in the resistor R_2 is

$$I_2 = \frac{R_0}{R_0 + R_2} I = \frac{R_0 E}{R_1(R_0 + R_2) + R_0 R_2} = \frac{R_0 E}{C R_0 + R_1 R_2}.$$

So that this current is at a minimum, $R_1 R_2 = R_1(C - R_1)$ must be at a maximum. Since this is rewritten as

$$R_1(C - R_1) = \frac{C^2}{4} - \left(\frac{C}{2} - R_1\right)^2,$$

the required condition is $R_1 = C/2$. Under this condition, $R_2 = C/2$ and we have

$$\frac{R_1}{R_2} = 1.$$

7.11   The voltage on the resistor R_1 is

$$V = \frac{R_1 E}{R_1 + R_2}$$

and the dissipated power is

$$P = \frac{V^2}{R_1} = \frac{R_1 E^2}{(R_1 + R_2)^2}.$$

So that P is at a maximum, the following condition must be satisfied:

$$\frac{\partial P}{\partial R_1} = \frac{(R_2 - R_1)E^2}{(R_1 + R_2)^3} = 0.$$

Thus, we have $R_1 = R_2$. In this case, we can prove $\partial^2 P/\partial R_1^2 < 0$, and the obtained condition is the solution. The dissipated power is

$$P = \frac{E^2}{4R_2}.$$

Chapter 8

8.1   The power is given by $P = E^2(t)/R$, and the energy dissipated in a period is determined to be

$$W = \int_{-T/2}^{0} \frac{E_0^2}{R}\left(1 + \frac{4t}{T}\right)^2 dt + \int_{0}^{T/2} \frac{E_0^2}{R}\left(1 - \frac{4t}{T}\right)^2 dt = \frac{E_0^2}{3R}T.$$

8.2   It is easier to solve for the electric charge Q. The equation for $t > 0$ is

$$R\frac{dQ}{dt} + \frac{Q}{C} = 0.$$

Since the electric charge $Q(0) = CE$ is stored in the capacitor in the initial condition, we assume the solution in the form: $Q(t) = CEe^{St}$. Then, we have $RS + C^{-1} = 0$. The solution for the electric charge is $Q(t) = CE\exp[-t/(CR)]$. The current is determined to be

$$I(t) = \frac{dQ(t)}{dt} = -\frac{E}{R}\exp\left(-\frac{t}{CR}\right).$$

The reason for a negative value of the current is that the current flows in the opposite direction (counterclockwise direction) to that in the process of charging the capacitor. The equation for the current is

$$RI(t) + \frac{1}{C}\left[CE + \int_0^t I(t)dt\right] = 0.$$

It can be easily shown that the above solution satisfies the equation.

8.3 Using $[R^2 - 4(L/C)]^{1/2}/(2L) = \beta$, we have $K_1 = -K_2 = -E/(2\beta L)$, and the following solution is obtained:

$$I(t) = -\frac{E}{2\beta L}\left\{\exp\left[-\left(\frac{R}{2L} - \beta\right)t\right] - \exp\left[-\left(\frac{R}{2L} + \beta\right)t\right]\right\}.$$

8.4 The electric charge $Q = CE$ is stored in the capacitor in the initial state, and the equation for $t > 0$ is given by

$$L\frac{dI(t)}{dt} + RI(t) + \frac{1}{C}\left[CE + \int_0^t I(t)dt\right] = 0.$$

We assume the solution of the form: $I(t) = Kt\,e^{St}$. Then, we have

$$K\left\{\left[S + \frac{2}{(LC)^{1/2}} + \frac{1}{SLC}\right]t + 1 - \frac{1}{LCS^2}\right\}e^{St} + \frac{K}{LCS^2} + \frac{E}{L} = 0.$$

From this equation S and K are determined, and the solution is given by

$$I(t) = -\frac{E}{L}t\exp\left[-\frac{t}{(LC)^{1/2}}\right].$$

8.5 The equation to solve is

$$L\frac{dI(t)}{dt} + RI(t) = 0.$$

We put $I(t) = I(0)\,e^{St}$. Then, from $LS + R = 0$ we have $S = -R/L$. Since the current is given by $I = E/R$ in the initial state, the current is determined to be

$$I(t) = \frac{E}{R}\exp\left(-\frac{R}{L}t\right).$$

The energy dissipated in the resistor until the steady state has been reached is

$$W = R\int_0^\infty I^2(t)dt = \frac{LE^2}{2R^2}.$$

This is the magnetic energy stored in the coil, $(1/2)LI^2(0)$, in the initial state, where $I(0) = E/R$ is the current in the initial state.

8.6 Using the solution to Exercise 8.2, the dissipated energy is

$$W = R\int_0^\infty I^2(t)dt = \frac{E^2}{R}\int_0^\infty \exp\left(-\frac{2}{CR}t\right)dt = \frac{1}{2}CE^2.$$

This is the electric energy stored in the capacitor in the initial state.

8.7 The dissipated energy is given by

$$W = R\int_0^\infty I^2(t)dt = \frac{E^2}{R}\int_0^\infty \exp\left(-\frac{2R}{L}t\right)dt = \frac{LE^2}{2R^2} = \frac{1}{2}LI^2(0),$$

where $I(0) = E/R$ is the current in the initial state. So, the energy dissipated in the resistor is the magnetic energy stored in the coil in the initial state.

8.8 Using the current given by Eq. (8.51), the energy stored in the coil is written as

$$W_L = \frac{1}{2}LI^2(t) = \frac{LE^2}{2Z_0^2}\cos^2(\omega t - \theta).$$

Since the electric charge stored in the capacitor is $Q(t) = [E/(\omega Z_0)]\sin(\omega t - \theta)$, the energy stored in the capacitor is

$$W_C = \frac{1}{2C}Q^2(t) = \frac{E^2}{2C\omega^2 Z_0^2}\sin^2(\omega t - \theta).$$

These energies vary sinusoidally with time, and there is no stored or dissipated energy during a cycle. The energy dissipated in the resistor during a cycle is

$$W_R = \int_0^{2\pi/\omega} RI^2(t)dt = \frac{RE^2}{Z_0^2}\int_0^{2\pi/\omega} \cos^2(\omega t - \theta)dt = \frac{\pi RE^2}{\omega Z_0^2}.$$

On the other hand, the energy supplied by the power source in a cycle is

$$W = \int_0^{2\pi/\omega} E(t)I(t)dt = \frac{E^2}{Z_0} \int_0^{2\pi/\omega} \cos\omega t \cos(\omega t - \theta)dt = \frac{\pi E^2}{\omega Z_0} \cos\theta = \frac{\pi R E^2}{\omega Z_0^2}.$$

Thus, all of the energy supplied by the power source is dissipated in the resistor and there is no energy newly stored in the coil and capacitor.

8.9 In the initial state $I(0) = E/R$. The equation that holds after the switch is turned is

$$L\frac{dI}{dt} + RI + \frac{1}{C}\int I dt = 0.$$

Here, we put as $I(t) = Ae^{St}$. Then, Eq. (8.24) is obtained from the above equation, and S is given by

$$S = -\alpha \pm i\beta; \quad \alpha = \frac{R}{2L}, \quad \beta = \frac{1}{2L}\left(\frac{4L}{C} - R^2\right)^{1/2}.$$

Then, the general solution is of the form:

$$I(t) = A_1 \exp[-(\alpha + i\beta)t] + A_2 \exp[-(\alpha - i\beta)t].$$

One of the conditions to be satisfied is the initial condition,

$$A_1 + A_2 = \frac{E}{R},$$

and the other is the condition that the electric charge stored in the capacitor is zero in the initial state:

$$\frac{A_1}{\alpha + i\beta} + \frac{A_2}{\alpha - i\beta} = 0.$$

From these conditions, we have

$$A_1 = \frac{E}{2R}\left(1 - i\frac{\alpha}{\beta}\right), \quad A_2 = \frac{E}{2R}\left(1 + i\frac{\alpha}{\beta}\right).$$

Taking the real part, the current is determined to be

$$I(t) = \frac{E}{R}e^{-\alpha t}\left(\cos\beta t - \frac{\alpha}{\beta}\sin\beta t\right)$$

$$= \frac{E}{R}\left(1 - \frac{CR^2}{4L}\right)^{-1} e^{-(R/2L)t} \cos\left[\frac{1}{2L}\left(\frac{4L}{C} - R^2\right)^{1/2} t + \theta\right],$$

where $\theta = \cot^{-1}\left[(4L/CR^2) - 1\right]^{1/2}$.

After a long calculation, we obtain the energy dissipated in the resistor until the steady state has been reached:

$$W = R \int_0^\infty I^2(t)\,dt = R\left[\frac{A_1^2}{2(\alpha + i\beta)} + \frac{A_1 A_2}{\alpha} + \frac{A_2^2}{2(\alpha - i\beta)}\right] = \frac{1}{2}L\left(\frac{E}{R}\right)^2.$$

This is rewritten as $(1/2)LI^2(0)$ in terms of the initial current value. Thus, this is the energy stored in the coil in the initial state.

8.10 The circuit equation is given by

$$RI + \frac{1}{C}\int I\,dt = E_0 \cos \omega t.$$

Here, we put the solution of the current in the steady state as

$$I(t) = K_1 \cos \omega t + K_2 \sin \omega t.$$

Then, the circuit equation is written as

$$\left(K_1 R - \frac{K_2}{\omega C}\right) \cos \omega t + \left(K_2 R + \frac{K_1}{\omega C}\right) \sin \omega t = E_0 \cos \omega t.$$

The conditions to be satisfied are:

$$K_1 = \frac{E_0 R}{R^2 + 1/(\omega C)^2}, \quad K_2 = -\frac{E_0}{\omega C\left[R^2 + 1/(\omega C)^2\right]}.$$

Thus, the current is determined to be

$$I(t) = \frac{E_0}{R^2 + 1/(\omega C)^2}\left(R \cos \omega t - \frac{1}{\omega C} \sin \omega t\right)$$

$$= \frac{E_0}{\left[R^2 + 1/(\omega C)^2\right]^{1/2}} \cos(\omega t + \theta),$$

where $\theta = \tan^{-1}(\omega C R)^{-1}$.

Chapter 9

9.1 We obtain $A_0 = 0$ from Eq. (9.3). We denote as $\omega t = \phi$. Then, we have

$$A_n = \frac{I_0}{\pi}\left(\int_0^\pi \cos n\phi \, d\phi - \int_\pi^{2\pi} \cos n\phi \, d\phi\right) = 0$$

for $n \geq 1$ and

$$B_n = \frac{I_0}{\pi}\left(\int_0^\pi \sin n\phi \, d\phi - \int_\pi^{2\pi} \sin n\phi \, d\phi\right) = \frac{2I_0}{n\pi}\left[1 - (-1)^n\right].$$

Thus, we obtain

$$I(t) = \sum_{m=0}^{\infty} \frac{4I_0}{(2m+1)\pi} \sin(2m+1)\omega t.$$

9.2 The impedance of the circuit is

$$Z = \frac{i\omega L R}{R + i\omega L} + \frac{1}{i\omega C} = \frac{(\omega L)^2 R}{R^2 + (\omega L)^2} + i\left[\frac{\omega L R^2}{R^2 + (\omega L)^2} - \frac{1}{\omega C}\right].$$

Thus, the resistance component and reactance are, respectively, given by

$$R_e = \frac{(\omega L)^2 R}{R^2 + (\omega L)^2}, \quad X = \frac{\omega L R^2}{R^2 + (\omega L)^2} - \frac{1}{\omega C}.$$

The admittance is

$$Y = \frac{\left[R^{-1} + (i\omega L)^{-1}\right] i\omega C}{R^{-1} + (i\omega L)^{-1} + i\omega C} = \frac{\omega C(\omega L - iR)}{R(\omega^2 LC - 1) - i\omega L}$$

$$= \frac{R\omega^4 L^2 C^2 + i\omega C\left[\omega^2 L^2 + R^2(1 - \omega^2 LC)\right]}{R^2(\omega^2 LC - 1)^2 + \omega^2 L^2}.$$

Thus, the conductance and susceptance are, respectively, given by

$$G = \frac{R\omega^4 L^2 C^2}{R^2(\omega^2 LC - 1)^2 + \omega^2 L^2}, \quad B = \frac{\omega C\left[\omega^2 L^2 + R^2(1 - \omega^2 LC)\right]}{R^2(\omega^2 LC - 1)^2 + \omega^2 L^2}.$$

9.3 From Eq. (9.27), we have

$$Z = Y^{-1} = \frac{1}{G + iB} = \frac{G - iB}{G^2 + B^2}.$$

Thus, the resistance component and reactance are, respectively, given by

$$R = \frac{G}{G^2 + B^2}, \quad X = -\frac{B}{G^2 + B^2}.$$

9.4 The impedance of the circuit is

$$Z = \frac{R_1 \omega L(\omega L + i R_1)}{R_1^2 + (\omega L)^2} + \frac{R_2(1 - i\omega C R_2)}{1 + (\omega C R_2)^2}$$

$$= \left[\frac{R_1(\omega L)^2}{R_1^2 + (\omega L)^2} + \frac{R_2}{1 + (\omega C R_2)^2} \right]$$

$$+ i\omega \left[\frac{R_1^2 L}{R_1^2 + (\omega L)^2} - \frac{R_2^2 C}{1 + (\omega C R_2)^2} \right].$$

Its real part is given by

$$\mathrm{Re}(Z) = \frac{R_1^2 R_2 + \omega^2 L^2 (R_1 + R_2) + \omega^4 (LCR_2)^2 R_1}{R_1^2 + \omega^2 \left(C^2 R_1^2 R_2^2 + L^2 \right) + \omega^4 (LCR_2)^2}.$$

So that this is independent of ω, the ratios of coefficients of ω^0, ω^2, and ω^4 must be the same between the denominator and numerator. From this condition we have

$$R_2 = \frac{L^2 (R_1 + R_2)}{C^2 R_1^2 R_2^2 + L^2} = R_1.$$

We representatively use R_1. The above equation leads to

$$L = CR_1^2. \tag{B.1}$$

Using the condition $R_2 = R_1$, the imaginary part of the impedance is written as

$$\mathrm{Im}(Z) = \omega \left[\frac{L}{1 + \omega^2 (L/R_1)^2} - \frac{R_1^2 C}{1 + \omega^2 (CR_1)^2} \right].$$

It is seen that the imaginary part is zero, if Eq. (B.1) is satisfied. Thus, the required condition is

$$R_1 = R_2 = \left(\frac{L}{C}\right)^{1/2}.$$

9.5 So that the bridge circuit is balanced, from Eq. (9.36) the following condition must be fulfilled:

$$R_2 R_3 = \left(R_1 + \frac{1}{i\omega C_1}\right)(R_4 + i\omega L_4).$$

Thus, we obtain

$$\frac{L_4}{C_1} = R_2 R_3 - R_1 R_4, \quad R_4 = \omega^2 C_1 R_1 L_4.$$

Then, R_4 and L_4 are determined to be

$$R_4 = \frac{\omega^2 C_1^2 R_1 R_2 R_3}{1 + \omega^2 C_1^2 R_1^2}, \quad L_4 = \frac{C_1 R_2 R_3}{1 + \omega^2 C_1^2 R_1^2}.$$

9.6 The impedance of the circuit is

$$Z = \frac{R}{1 + \omega^2 C^2 R^2} + i\omega\left(L - \frac{CR^2}{1 + \omega^2 C^2 R^2}\right).$$

So, the requirement is satisfied if the imaginary part is zero. Thus, we have

$$\omega = \left(\frac{CR^2 - L}{C^2 R^2 L}\right)^{1/2}.$$

9.7 The voltage on the resistor and coil is

$$V = E\frac{R + i\omega L}{R + i[\omega L - 1/(\omega C)]}$$
$$= \frac{E(R + i\omega L)\{R - i[\omega L - 1/(\omega C)]\}}{R^2 + [\omega L - 1/(\omega C)]^2}.$$

The numerator is proportional to

$$R^2 + (\omega L)^2 - \frac{L}{C} + i\frac{R}{\omega C}.$$

Hence, if the real part is the same as the imaginary part, the required condition is satisfied:

$$\frac{1}{C}\left(L + \frac{R}{\omega}\right) = R^2 + (\omega L)^2.$$

Thus, we obtain the capacitance

$$C = \frac{R + \omega L}{\omega [R^2 + (\omega L)^2]}.$$

9.8 The relation $V_1/V = Z/(R + iX + Z)$ is written as

$$\left| \frac{V_1}{V} \right| = \frac{|Z|}{[(R + |Z| \cos \varphi)^2 + (X + |Z| \sin \varphi)^2]^{1/2}}.$$

So that this is at a minimum, the following quantity in the denominator must be at a maximum:

$$(R + |Z| \cos \varphi)^2 + (X + |Z| \sin \varphi)^2 = R^2 + X^2 + |Z|^2 + 2|Z|(R \cos \varphi + X \sin \varphi).$$

From the condition that the differentiation of this quantity with respect to φ is zero, we have $-R \sin \varphi + X \cos \varphi = 0$. This leads to

$$\varphi = \tan^{-1} \frac{X}{R}.$$

In this case, the second derivative with respect to φ is

$$-(R \cos \varphi + X \sin \varphi) = -R \cos \varphi \left(1 + \frac{X^2}{R^2}\right) < 0,$$

since $R > 0$ and $\cos \varphi > 0$. Thus, the denominator is at a maximum, and the above condition for φ is the solution.

9.9 The total impedance is

$$Z = i\omega L + \frac{R_1[R_2 + 1/(i\omega C)]}{R_1 + R_2 + 1/(i\omega C)}.$$

The current I_2 flowing in the resistor R_2 is obtained from the total current $I = E/Z$, and the voltage of the resistor is

$$V = \frac{E}{Z} \cdot \frac{R_1}{R_1 + R_2 + 1/(i\omega C)} \cdot R_2$$

$$= \frac{R_1 R_2 E}{R_1 R_2 + (L/C) + i[\omega L(R_1 + R_2) - R_1/(\omega C)]}.$$

This leads to

$$|V| = \frac{R_1 R_2 |E|}{\{[R_1 R_2 + (L/C)]^2 + [\omega L(R_1 + R_2) - R_1/(\omega C)]^2\}^{1/2}}.$$

So that $|V|$ is at a maximum, the following condition must be satisfied:

$$\omega L(R_1 + R_2) - \frac{R_1}{\omega C} = 0.$$

Thus, we have

$$\omega^2 = \frac{R_1}{LC(R_1 + R_2)}.$$

9.10 The ratio of the input and output voltages for the low-pass filter shown in Fig. 9.20 is

$$\frac{V_2}{V_1} = \frac{R[R(1 - \omega^2 LC) - i\omega L]}{R^2(1 - \omega^2 LC)^2 + \omega^2 L^2},$$

and its phase is

$$\varphi = \arg\left(\frac{V_2}{V_1}\right) = -\tan^{-1}\frac{\omega L}{R(1 - \omega^2 LC)}.$$

In terms of the cut-off angular frequency, $\omega_c = 1/(LC)^{1/2}$, the phase at $R = \omega_c L/\sqrt{2}$ is

$$\varphi = -\tan^{-1}\frac{\sqrt{2}(\omega/\omega_c)}{1 - (\omega/\omega_c)^2}.$$

The angular frequency dependence of the phase is the same as that for the parallel resonance circuit shown in Fig. 9.16. This is shown in Fig. 9.18.

Next, the ratio of the input and output voltages for the high-pass filter shown in Fig. 9.22 is

$$\frac{V_2}{V_1} = -\frac{\omega^2 LCR[R(1 - \omega^2 LC) - i\omega L]}{R^2(1 - \omega^2 LC)^2 + \omega^2 L^2},$$

and its phase is

$$\varphi = \tan^{-1}\frac{\sqrt{2}(\omega/\omega_c)}{1 - (\omega/\omega_c)^2}.$$

Thus, the phase is opposite to that of the parallel resonance circuit and the low-path filter.

Fig. B.11 Angular frequency dependence of notch filter for $\Delta\omega/\omega_c = 0.01$

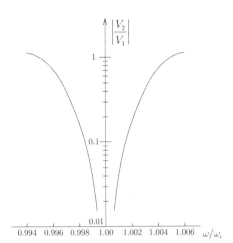

9.11   The ratio of the input and output voltages is given by

$$\frac{V_2}{V_1} = \left[1 - \frac{\omega^2 L_1 C_2 R + i\omega L_1 (\omega^2 L_2 C_2 - 1)}{R(\omega^2 L_1 C_1 - 1)(\omega^2 L_2 C_2 - 1)}\right]^{-1}.$$

Substituting the conditions for the capacitances and inductances, we have

$$\left|\frac{V_2}{V_1}\right| = \left[1 - \frac{\omega^2 \Delta\omega^2}{(\omega^2 - \omega_c^2)^2} + \frac{\omega^4 \Delta\omega^4}{(\omega^2 - \omega_c^2)^4}\right]^{-1/2}.$$

It can be shown that $|V_2/V_1| = 0$ at $\omega = \omega_c$ and $|V_2/V_1| = 1/\sqrt{3}$ at $\omega = \omega_c \pm \Delta\omega/2\sqrt{2}$. Figure B.11 shows the angular frequency dependence of $|V_2/V_1|$ for the case of $\Delta\omega/\omega_c = 0.01$.

9.12   The impedance of the circuit is

$$Z = \frac{-3\omega^2 L^2 + 2i\omega L R}{R + 2i\omega L}$$

and the current is given by

$$I = \frac{(R + 2i\omega L)V}{-3\omega^2 L^2 + 2i\omega L R} = \frac{[\omega L R - 2i(R^2 + 3\omega^2 L^2)]V}{\omega L(4R^2 + 9\omega^2 L^2)}.$$

So, the dissipated power is determined to be

$$P = \mathrm{Re}(V^* I) = \frac{R|V|^2}{4R^2 + 9\omega^2 L^2}.$$

From the condition $\partial P/\partial R = 0$, we have $9\omega^2 L^2 - 4R^2 = 0$, which leads to

$$R = \frac{3}{2}\omega L.$$

Since the second derivative of P with respect to R is negative, the dissipated power is at a maximum. Hence, the above condition for R is the solution.

9.13 (a) The conditions among the currents are $I_1 = I_2 + I_3$ and $RI_2 = ZI_3$. The second condition yields

$$|I_2|^2 = \frac{|Z|^2}{R^2}|I_3|^2.$$

From the condition $I_1 = [1 + (Z/R)]I_3$, we have

$$|I_1|^2 = \left(1 + \frac{Z}{R}\right)\left(1 + \frac{Z^*}{R}\right)|I_3|^2 = \left[1 + \frac{1}{R}(Z + Z^*) + \frac{|Z|^2}{R^2}\right]|I_3|^2.$$

(b) Since the resistance component of impedance Z is given by $(Z + Z^*)/2$, the dissipated power is

$$P = \frac{1}{2}(Z + Z^*)|I_3|^2.$$

(c) Using the above results, we have

$$|I_1|^2 - |I_2|^2 - |I_3|^2 = \left[1 + \frac{1}{R}(Z + Z^*) + \frac{|Z|^2}{R^2} - \frac{|Z|^2}{R^2} - 1\right]|I_3|^2$$

$$= \frac{1}{R}(Z + Z^*)|I_3|^2.$$

Thus, the proposition is proved:

$$\frac{1}{2}R(|I_1|^2 - |I_2|^2 - |I_3|^2) = \frac{1}{2}(Z + Z^*)|I_3|^2 = P.$$

Chapter 10

10.1 The stored energy in the transformer circuit shown in Fig. 10.1 is given by Eq. (10.14), which is rewritten as

$$U_m = \frac{1}{2}L_1\left(I_1 + \frac{M}{L_1}I_2\right)^2 + \frac{1}{2}\left(L_2 - \frac{M^2}{L_1}\right)I_2^2.$$

It can happen that the first term is zero. Hence, so that the energy is not negative, the condition, $L_2 - M^2/L_1 \geq 0$, must be satisfied. Thus, Eq. (10.3) is derived.

Fig. B.12 Equivalent circuit

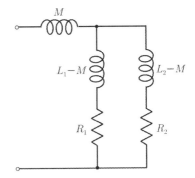

10.2 The equivalent circuit is shown in Fig. B.12. Hence, the impedance is given by

$$Z = i\omega M + \frac{[R_1 + i\omega(L_1 - M)][R_2 + i\omega(L_2 - M)]}{R_1 + R_2 + i\omega(L_1 + L_2 - 2M)}$$

$$= \frac{R_1 R_2 - \omega^2(L_1 L_2 - M^2) + i\omega(R_1 L_2 + R_2 L_1)}{R_1 + R_2 + i\omega(L_1 + L_2 - 2M)}.$$

10.3 From the second equation in the solution to Example 10.1, we have

$$\left[R + i\left(\omega L_2 - \frac{1}{\omega C}\right)\right]I_2 = -i\left(\omega M - \frac{1}{\omega C}\right)I_1.$$

Hence, so that $I_2 = 0$ is fulfilled, $\omega M - 1/(\omega C) = 0$ must be satisfied. That is,

$$\omega = \left(\frac{1}{MC}\right)^{1/2}.$$

10.4 The equivalent circuit is shown in Fig. B.13. Hence, the potential difference between A and B becomes zero, when the following condition is satisfied:

$$R_1[R_4 + i\omega(L_4 + M)] = [R_2 + i\omega(L_2 - M)]R_3.$$

Thus, R_4 and L_4 are obtained as

$$R_4 = \frac{R_2 R_3}{R_1}, \quad L_4 = \frac{R_3}{R_1}(L_2 - M) - M.$$

Fig. B.13 Equivalent circuit

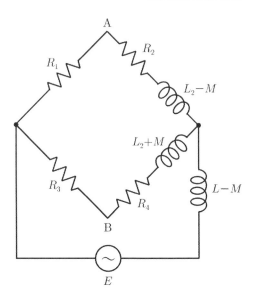

10.5 The circuit equation is

$$i\omega L_1 I_1 + i\omega M I_2 = i\omega L_2 I_2 + i\omega M I_1 + R I_2.$$

Then, we have

$$\frac{I_1}{I_2} = \frac{i\omega(L_2 - M) + R}{i\omega(L_1 - M)}.$$

So that the required condition is realized, $M = L_2$ and $\omega(L_1 - M) = \pm R$ must be satisfied. The second condition is rewritten as $M = L_1 \mp R/\omega$. Because $M^2 \leq L_1 L_2$, the second condition must be $M = L_1 - R/\omega$. Thus, we have

$$\omega = \frac{R}{L_1 - M} = \frac{R}{L_1 - L_2}.$$

Chapter 11

11.1 Equation (11.1) is written as

$$-I_1 + I_3 + I_4 = 0, \quad -I_4 - I_5 + I_6 = 0,$$
$$-I_2 - I_3 + I_5 = 0, \quad I_1 + I_2 - I_6 = 0,$$

for four nodes. Since there are three independent relations for six branch currents, the number of independent branch currents are three. If we choose I_3, I_4, and I_5 for the independent variables, other branch currents are given by

$$I_1 = I_3 + I_4, \quad I_2 = -I_3 + I_5, \quad I_6 = I_4 + I_5.$$

Equation (11.2) for each closed path is given by

$$E_1 = R_1 I_1 + R_4 I_4 + R_6 I_6 = R_1 I_3 + (R_1 + R_4 + R_6) I_4 + R_6 I_5,$$
$$E_2 = R_2 I_2 + R_5 I_5 + R_6 I_6 = -R_2 I_3 + R_6 I_4 + (R_2 + R_5 + R_6) I_5,$$
$$E_3 = R_3 I_3 - R_4 I_4 + R_5 I_5.$$

The currents I_3, I_4, and I_5 are determined from these conditions.

11.2 Since all branch currents can be expressed using three closed currents, the number of independent branch currents is three. Among them, I_3 is given by J_3, and the number of remaining independent branch currents is two. We use I_5 and I_6 for them. Then, the other currents are given by

$$I_1 = J_3 + I_5 + I_6, \quad I_2 = J_3 + I_5, \quad I_4 = I_5 + I_6. \tag{B.2}$$

We adopt two closed paths for the conditions to determine the two currents. Equation (11.2) becomes

$$E_1 = R_1 J_3 + (R_1 + R_4) I_5 + (R_1 + R_4 + R_6) I_6$$

for the closed path composed of R_1, R_4, and R_6, and

$$E_2 = R_2 J_3 + (R_2 + R_5) I_5 - R_6 I_6$$

for the closed path composed of R_2, R_5, and R_6. From these equations we obtain

$$I_6 = \frac{(R_2 + R_5) E_1 - (R_1 + R_4) E_2 - (R_1 R_5 - R_2 R_4) J_3}{(R_1 + R_4 + R_6)(R_2 + R_5 + R_6) - R_6^2}$$

and

$$I_5 = \frac{R_6 E_1 + (R_1 + R_4 + R_6) E_2 - [R_1 R_6 + R_2(R_1 + R_4 + R_6)] J_3}{(R_1 + R_4 + R_6)(R_2 + R_5 + R_6) - R_6^2}.$$

Each branch currents are obtained by substitution of these currents into Eq. (B.2).

To realize $I_6 = 0$, the condition, $(E + R J_3)(R_2 - R_1) = 0$, must be satisfied. Thus, we obtain $R_1 = R_2$.

11.3 Closed currents are defined, as shown in Fig. B.14. Equation (11.2) is written as follows for closed currents I_a, I_b, and I_c:

$$0 = (R + R_1 + R_3) I_a - R I_b - R_3 I_c,$$

Fig. B.14 Closed currents

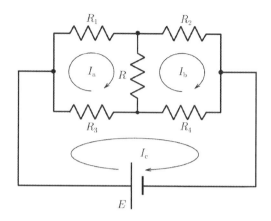

$$0 = -RI_a + (R + R_2 + R_4)I_b - R_4 I_c,$$
$$E = -R_3 I_a - R_4 I_b + (R_3 + R_4)I_c.$$

These closed currents are obtained using the above equations.

11.4 The following equation holds for the left closed path including R_3:

$$E_1 = R_3 I_3.$$

Thus, we have $I_3 = E_1/R_3$. The equation for the right closed path is

$$E_2 = R_2 I_2 + R_3 I_3.$$

This yields $I_2 = (E_2 - E_1)/R_2$. Then, I_1 is determined to be

$$I_1 = I_3 - I_2 = \frac{E_1}{R_3} - \frac{E_2 - E_1}{R_2} = \frac{(R_2 + R_3)E_1 - R_3 E_2}{R_2 R_3}.$$

This agrees with Eq. (11.19).

11.5 We can use I_1 for the closed current for the left closed path including R_3. The closed current for the right closed path including R_3 is denoted by I' in the direction of branch current I_3. The following equations hold for these closed paths:

$$E_1 = R_3(I_1 + I'), \quad E_2 = R_3(I_1 + I') + R_2 I'.$$

Thus, we have $I' = (E_2 - E_1)/R_2$, and I_1 is determined to be

$$I_1 = \frac{E_1}{R_3} - I' = \frac{E_1}{R_3} - \frac{E_2 - E_1}{R_2} = \frac{(R_2 + R_3)E_1 - R_3 E_2}{R_2 R_3}.$$

Fig. B.15 Closed currents

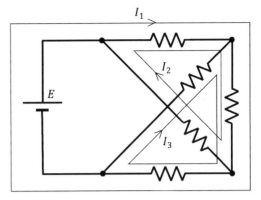

11.6 Closed currents I_1, I_2, and I_3 are defined as shown in Fig. B.15. The following equations hold for these closed paths:

$$E = R(3I_1 + 2I_2 + 2I_3),$$
$$0 = R(2I_1 + 3I_2 + I_3),$$
$$0 = R(2I_1 + I_2 + 3I_3).$$

Eliminating I_1 from the second and third equations, we have $I_2 = I_3 = -I_1/2$. Substituting this into the first equation, the current is determined to be

$$I = I_1 = \frac{E}{R}.$$

11.7 The equivalent circuit is shown in Fig. B.16. The equations for closed currents I_1, I_2, and I_3 are given by

Fig. B.16 Equivalent circuit

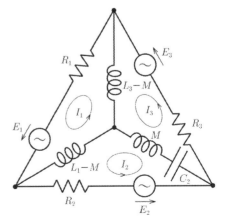

$$E_1 = R_1 I_1 + i\omega(L_1 - M)(I_1 - I_2) + i\omega(L_3 - M)(I_1 - I_3)$$
$$= [R_1 + i\omega(L_1 + L_3 - 2M)]I_1 - i\omega(L_1 - M)I_2 - i\omega(L_3 - M)I_3,$$

$$E_2 = R_2 I_2 + \left(i\omega M + \frac{1}{i\omega C_2}\right)(I_2 - I_3) + i\omega(L_1 - M)(I_2 - I_1)$$
$$= -i\omega(L_1 - M)I_1 + \left(R_2 + i\omega L_1 + \frac{1}{i\omega C_2}\right)I_2 - \left(i\omega M + \frac{1}{i\omega C_2}\right)I_3,$$

$$E_3 = R_3 I_3 + i\omega(L_3 - M)(I_3 - I_1) + \left(i\omega M + \frac{1}{i\omega C_2}\right)(I_3 - I_2)$$
$$= -i\omega(L_3 - M)I_1 - \left(i\omega M + \frac{1}{i\omega C_2}\right)I_2 + \left(R_3 + i\omega L_3 + \frac{1}{i\omega C_2}\right)I_3.$$

11.8   When only voltage source E_1 remains, the current flowing in R_2 is

$$I_1 = \frac{R_3 E_1}{R_1 R_2 + R_2 R_3 + R_3 R_1}.$$

When only current source J_2 remains, the current flowing in R_2 is

$$I_2 = \frac{R_1 R_3 J_2}{R_1 R_2 + R_2 R_3 + R_3 R_1}.$$

When only voltage source E_3 remains, the current flowing in R_2 is

$$I_3 = \frac{R_1 E_3}{R_1 R_2 + R_2 R_3 + R_3 R_1}.$$

From these results, the current I is determined to be

$$I = I_1 + I_2 + I_3 = \frac{R_3 E_1 + R_1 R_3 J_2 + R_1 E_3}{R_1 R_2 + R_2 R_3 + R_3 R_1}.$$

11.9   When DC current source of 1 A is connected between P and infinity, current
       1/4 A flows directly from P to Q from symmetry. When DC current source of
       1 A is connected between infinity and Q, current 1/4 A flows directly from P
       to Q from symmetry. From superposition of these cases, current 1/2 A flows
       from P to Q. Thus, the voltage between P and Q is $R/2$ V, and the resistance
       is determined to be $R/2$ Ω.

11.10 When resistance R is removed, the open-circuited voltage is obtained by
       superposing the contribution from the voltage source, $R_2 E_1/(R_1 + R_2)$, and
       that of the current source, $R_1 R_2 J_2/(R_1 + R_2)$:

$$V_0 = \frac{R_2(E_1 + R_1 J_2)}{R_1 + R_2}.$$

When the effects of the two power sources are removed, the resistance is $R_0 = R_1 R_2/(R_1 + R_2)$. So, when resistance R is connected, the current that flows in R is given by

$$I = \frac{V_0}{R_0 + R} = \frac{R_2(E_1 + R_1 J_2)}{R_1 R_2 + R(R_1 + R_2)}.$$

11.11   The open-circuited voltage due to the voltage source is $aE/(1 + a)$ and that due to the current source is $aRJ/(1 + a)$ in Fig. 11.32a. So, superposing these, the open-circuited voltage is given by $a(E + RJ)/(1 + a)$. Comparing this with the equivalent circuit in Fig. 11.32b, the following must be satisfied:

$$\frac{a(E + RJ)}{1 + a} = E.$$

When the voltage source is short-circuited and the current source is removed and opened, the resistance between the terminals is $aR/(1 + a)$. This must be equal to $2R/3$. Thus, we have $a = 2$. Then, from the above equation we have $J = E/(2R)$.

11.12   When the terminals a and b are short-circuited in Fig. 11.25, the current is given by the sum of currents $Y_i E_i$ from each power source, as shown by the principle of superposition:

$$I_0 = \sum_{i=1}^{n} Y_i E_i.$$

When all voltage sources are short-circuited, the admittance between the terminals a and b is

$$Y_0 = \sum_{i=1}^{n} Y_i.$$

Thus, when the terminals a and b are opened ($Y = 0$), the voltage between the terminals is obtained from Eqs. (11.37), and (11.38) is derived.

Index